CALCULUS II
Guided Notebook

THIRD EDITION

JOHN R. TAYLOR • DESIRÉ J. TAYLOR

Kendall Hunt
publishing company

Cover image © Kuttelvaserova Stuchelova/Shutterstock.com

All images © Kendall Hunt Publishing unless otherwise noted

www.kendallhunt.com
Send all inquiries to:
4050 Westmark Drive
Dubuque, IA 52004-1840

Copyright © 2020, 2021, 2022 by John Taylor and Desiré Taylor

Student version ISBN: 979-8-7657-0745-6
Faculty version ISBN: 979-8-7657-0746-3

Kendall Hunt Publishing Company has the exclusive rights to reproduce this work, to prepare derivative works from this work, to publicly distribute this work, to publicly perform this work and to publicly display this work.

All rights reserved. No part of this publication may be reproduced, stored in a retrieval system, or transmitted, in any form or by any means, electronic, mechanical, photocopying, recording, or otherwise, without the prior written permission of the copyright owner.

Published in the United States of America

Contents

Section	Topic	Page
5R	Calculus I Review—Differentiation and Limits	1
5.1	Areas and Distances (Riemann Sums)	3
5.2	The Definite Integral	11
5.3	Evaluating Integrals	19
5.4	The Fundamental Theorem of Calculus	29
5.5	Integration by U-Substitution	37
6.1	Integration by Parts	47
6.3	Partial Fractions and Additional Techniques of Integration	57
6.4	Integration with Tables	75
6.5	Approximation	89
6.6	Improper Integrals	99
7.1	Area Between Curves	113
7.2/7.3	Volume	123
7.4	Arc Length	149
7.6	Work	155
8.1	Sequence	177
8.2	Series	187
8.3	Integral and Comparison Tests	199
8.4	Alternating Series Test	211
8.5	Power Series	227
8.6	Representation of Functions as a Power Series	235
8.7	Taylor and MacLauren Series	245

Calculus I Review—Differentiation and Limits

Section 5R

A. Properties and Formulas of Differentiation

1. Basic Functions

1. $\dfrac{d}{dx}(c) = 0$

2. $\dfrac{d}{dx}[c \cdot f(x)] = c \cdot f'(x)$

3. $\dfrac{d}{dx}[f(x) + g(x)] = f'(x) + g'(x)$

4. $\dfrac{d}{dx}[f(x) - g(x)] = f'(x) - g'(x)$

5. $\dfrac{d}{dx}(x^n) = n \cdot x^{n-1}$

6. $\dfrac{d}{dx}[f(x) \cdot g(x)] = f'(x)g(x) + f(x)g'(x)$

7. $\dfrac{d}{dx}\left[\dfrac{f(x)}{g(x)}\right] = \dfrac{f'(x)g(x) - f(x)g'(x)}{[g(x)]^2}$

8. $\dfrac{d}{dx}f(g(x)) = f'(g(x)) \cdot g'(x)$

2. Logarithmic and Exponential Functions

1. $\dfrac{d}{dx}(e^x) = e^x$

2. $\dfrac{d}{dx}(a^x) = a^x \cdot \ln a$

3. $\dfrac{d}{dx}\ln|x| = \dfrac{1}{x}$

4. $\dfrac{d}{dx}\log_a x = \dfrac{1}{x \cdot \ln a}$

3. Trigonometric Functions

1. $\dfrac{d}{dx}(\sin x) = \cos x$

2. $\dfrac{d}{dx}(\cos x) = -\sin x$

3. $\dfrac{d}{dx}(\tan x) = \sec^2 x$

4. $\dfrac{d}{dx}(\csc x) = -\csc x \cdot \cot x$

5. $\dfrac{d}{dx}(\sec x) = \sec x \cdot \tan x$

6. $\dfrac{d}{dx}(\cot x) = -\csc^2 x$

4. Inverse Trigonometric Functions

1. $\dfrac{d}{dx}\left(\sin^{-1} x\right) = \dfrac{1}{\sqrt{1-x^2}}$
2. $\dfrac{d}{dx}\left(\cos^{-1} x\right) = -\dfrac{1}{\sqrt{1-x^2}}$
3. $\dfrac{d}{dx}\left(\tan^{-1} x\right) = \dfrac{1}{1+x^2}$
4. $\dfrac{d}{dx}\left(\csc^{-1} x\right) = -\dfrac{1}{x\cdot\sqrt{1-x^2}}$
5. $\dfrac{d}{dx}\left(\sec^{-1} x\right) = \dfrac{1}{x\cdot\sqrt{1-x^2}}$
6. $\dfrac{d}{dx}\left(\cot^{-1} x\right) = -\dfrac{1}{1+x^2}$

B. Basic Limit Properties and Techniques

1. *Theorem:* We say that a limit exists when the limit from the left equals the limit from the right.
$$\lim_{x\to a} f(x) = L \iff \lim_{x\to a^-} f(x) = L = \lim_{x\to a^+} f(x)$$

2. *Direct Substitution Property:* If f is a polynomial or rational function and $a \in$ Domain, then $\lim_{x\to a} f(x) = f(a)$
Example: $\lim_{x\to 1} x^2 + 2x = (1)^2 + 2(1) = 3$

3. *Factoring/Manipulation (then Evaluation):* Factor expressions and cancel any common terms.
Example: $\lim_{x\to 4} \dfrac{3x-12}{x^2-16} = \dfrac{3(x-4)}{(x+4)(x-4)} = \dfrac{3}{(x+4)} = \dfrac{3}{(4)+4} = \dfrac{3}{8}$

4. *Indeterminate Form:* If $\lim_{x\to a} f(x) = f(a) = \dfrac{0}{0}$, then **factor, simplify,** or multiply by **conjugate**.
Example: $\lim_{x\to 4} \dfrac{x^2-16}{x-4} = \dfrac{0}{0} \Rightarrow \lim_{x\to 4} \dfrac{x^2-16}{x-4} = \dfrac{2x-4}{1} = 4$

5. *Undefined Form:* If $\lim_{x\to a} f(x) = f(a) = \dfrac{\text{Number}}{0}$, then the limit does not exist—DNE.
Example: $\lim_{x\to 1} \dfrac{x^2+2x+1}{x-1} = \dfrac{3}{0} \Rightarrow$ DNE

6. *Limits as Infinity:* For positive integers M and N such that $M > N$
   i. Degree of the Numerator = Degree of the Denominator
   $$\lim_{x\to\infty} \dfrac{\text{Polynomail of Degree } M}{\text{Polynomail of Degree } M} = \text{Ratio of Leading Coeficients}$$
   ii. Degree of the Numerator > Degree of the Denominator
   $$\lim_{x\to\infty} \dfrac{\text{Polynomail of Degree } M}{\text{Polynomail of Degree } N} = \pm\infty$$
   iii. Degree of the Numerator < Degree of the Denominator
   $$\lim_{x\to\infty} \dfrac{\text{Polynomail of Degree } N}{\text{Polynomail of Degree } M} = 0$$

7. *L'Hospital's Rule:* Suppose that $f(x)$ and $g(x)$ are differentiable, $g'(x) \neq 0$ and that
$$\lim_{x\to a} \dfrac{f(x)}{g(x)} = \dfrac{0}{0} \text{ or that } \lim_{x\to a} \dfrac{f(x)}{g(x)} = \dfrac{\infty}{\infty} \text{ then } \lim_{x\to a} \dfrac{f(x)}{g(x)} = \dfrac{f'(x)}{g'(x)}$$

Areas and Distances (Riemann Sums)

Section 5.1

Before Class Video Examples

1. If you are asked to find the Left Riemann Sum for $f(x) = x^2 + 3$ over the interval from $[0, 2]$ with $n = 4$

   a. Calculate $\Delta x =$

   b. Calculate the Left Riemann Sum

   c. Calculate the Right Riemann Sum

4   Section 5.1: Areas and Distances (Riemann Sums)

2. The speed of a runner increased steadily during the first 2 seconds of a race.

t(s)	0	0.5	1	1.5	2
v(ft/s)	0	5.1	9.3	12.5	15.6

Estimate the distance he or she traveled during these 2 seconds using the speed at the end of the time intervals.

Algebra Review

1. **Evaluating Functions**

    Example

    i. For the function $f(x) = 3x^2 + 4$, find $f(0)$ and $f\left(\dfrac{1}{2}\right)$

2. **Area of Rectangles**

    $$\text{Area} = \text{Length} \times \text{Width}$$

    Example

    ii. Give the area of the rectangle

    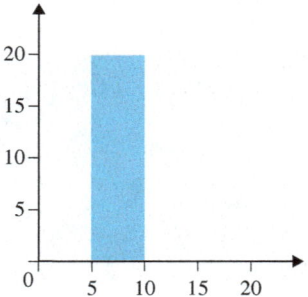

Section 5.1: Areas and Distances (Riemann Sums)

A. Riemann Sums

Riemann Sums: A technique of estimating the area under a curve by dividing the area into rectangles.

Examples
1. Consider the graph of $y = x^2$. Estimate the area under the curve between $x = 0$ and $x = 2$.

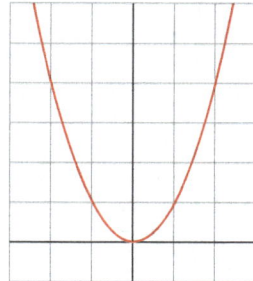

Divide the area up into four rectangles and find the area of each. Note that the interval $(0, 2)$ has been divided into four equal parts—each with width $\frac{1}{2}$.

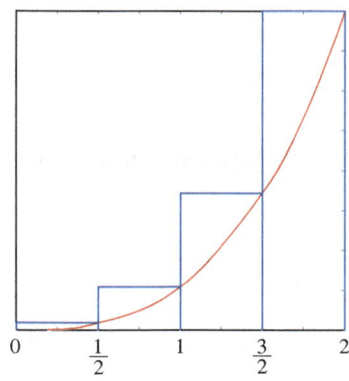

The height (length) of each rectangle can be found by calculating the function value at the points
$x = \frac{1}{2}$, $x = 1$, $x = \frac{3}{2}$, and $x = 2$.
Then the areas respectively are the following:

$$A_1 = l \cdot w = f\left(\frac{1}{2}\right) \cdot \left(\frac{1}{2}\right) = \left(\frac{1}{2}\right)^2 \cdot \left(\frac{1}{2}\right) = \left(\frac{1}{4}\right) \cdot \left(\frac{1}{2}\right) = \frac{1}{8}$$

$$A_2 = l \cdot w = f(1) \cdot \left(\frac{1}{2}\right) = (1)^2 \cdot \left(\frac{1}{2}\right) = (1) \cdot \left(\frac{1}{2}\right) = \frac{1}{2}$$

$$A_3 = l \cdot w = f\left(\frac{3}{2}\right) \cdot \left(\frac{1}{2}\right) = \left(\frac{3}{2}\right)^2 \cdot \left(\frac{1}{2}\right) = \left(\frac{9}{4}\right) \cdot \left(\frac{1}{2}\right) = \frac{9}{8}$$

$$A_4 = l \cdot w = f(2) \cdot \left(\frac{1}{2}\right) = (2)^2 \cdot \left(\frac{1}{2}\right) = (4) \cdot \left(\frac{1}{2}\right) = 2$$

Total Area: $A_1 + A_2 + A_3 + A_4 = \frac{1}{8} + \frac{1}{2} + \frac{9}{8} + 2 = \frac{30}{8}$

Since we used the right-hand point of each interval, we call this a **Right Riemann Sum**. Since this is an increasing function on the interval, it yielded an overestimate (called the Upper Riemann Sum).

Section 5.1: Areas and Distances (Riemann Sums)

2. Calculate the **Left Riemann Sum** (Lower Riemann Sum).

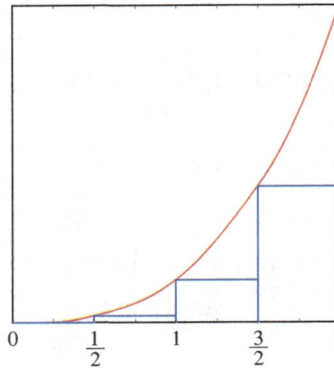

The height (length) of each rectangle can be found by calculating the function value at the points

$x = 0$, $x = \dfrac{1}{2}$, $x = 1$, and $x = \dfrac{3}{2}$.

The areas respectively are the following:

$A_0 = l \cdot w = f(0) \cdot \left(\dfrac{1}{2}\right) = (0)^2 \cdot \left(\dfrac{1}{2}\right) = (0) \cdot \left(\dfrac{1}{2}\right) = 0$

$A_1 = l \cdot w = f\left(\dfrac{1}{2}\right) \cdot \left(\dfrac{1}{2}\right) = \left(\dfrac{1}{2}\right)^2 \cdot \left(\dfrac{1}{2}\right) = \left(\dfrac{1}{4}\right) \cdot \left(\dfrac{1}{2}\right) = \dfrac{1}{8}$

$A_2 = l \cdot w = f(1) \cdot \left(\dfrac{1}{2}\right) = (1)^2 \cdot \left(\dfrac{1}{2}\right) = (1) \cdot \left(\dfrac{1}{2}\right) = \dfrac{1}{2}$

$A_3 = l \cdot w = f\left(\dfrac{3}{2}\right) \cdot \left(\dfrac{1}{2}\right) = \left(\dfrac{3}{2}\right)^2 \cdot \left(\dfrac{1}{2}\right) = \left(\dfrac{9}{4}\right) \cdot \left(\dfrac{1}{2}\right) = \dfrac{9}{8}$

Total Area: $A_0 + A_1 + A_2 + A_3 = 0 + \dfrac{1}{8} + \dfrac{1}{2} + \dfrac{9}{8} = \dfrac{14}{8}$

A better estimate is the average of the two:

$\dfrac{\text{Left Riemann Sum} + \text{Right Riemann Sum}}{2} = \dfrac{\left(\dfrac{30}{8}\right) + \left(\dfrac{14}{8}\right)}{2} = \dfrac{11}{4}$

3. We can repeat the process using the midpoint of the interval to find the "**Midpoint Riemann Sum**" or just called the Midpoint Rule.

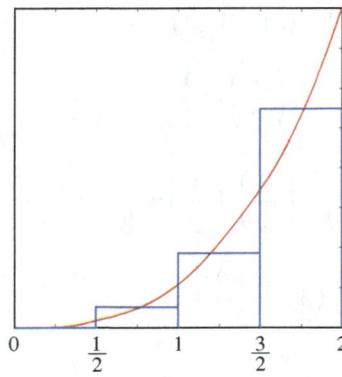

The height (length) of each rectangle can be found by calculating the function value at the points.

The areas respectively are the following:

$A_1 = l \cdot w =$

$A_2 = l \cdot w =$

$A_3 = l \cdot w =$

$A_4 = l \cdot w =$

B. Riemann Formulas

For an interval (a,b) that is divided into n subintervals, each with length $\Delta x = \dfrac{b-a}{n}$

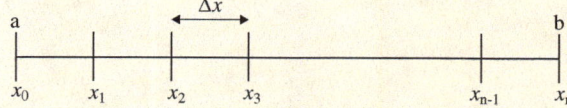

The area given by the

Left Riemann Sum is
$$A = f(x_0) \cdot \Delta x + f(x_1) \cdot \Delta x + \ldots + f(x_{n-1}) \cdot \Delta x = \sum_{i=0}^{n-1} f(x_i) \cdot \Delta x$$

Right Riemann Sum is
$$A = f(x_1) \cdot \Delta x + f(x_2) \cdot \Delta x + \ldots + f(x_n) \cdot \Delta x = \sum_{i=1}^{n} f(x_i) \cdot \Delta x$$

Midpoint Riemann Sum is
$$A = f\left(\frac{x_0+x_1}{2}\right) \cdot \Delta x + f\left(\frac{x_1+x_2}{2}\right) \cdot \Delta x + \ldots + f\left(\frac{x_{n-1}+x_n}{2}\right) \cdot \Delta x = \sum_{i=1}^{n} f(\overline{x}) \cdot \Delta x$$

More Examples

4. For the function $f(x) = 3x^2 + 2$ on $[0, 2]$ and $n = 4$,

   a. Find the Left Riemann Sum

   b. Find the Right Riemann Sum

   c. Find the Midpoint Riemann Sum

Section 5.1: Areas and Distances (Riemann Sums)

5. Estimate the area under the graph of $f(x) = |9 - x|$ from $x = 7$ to $x = 11$ using the midpoint rule with $n = 4$ (round your answer to six decimal places).

C. Accuracy in Riemann Sums

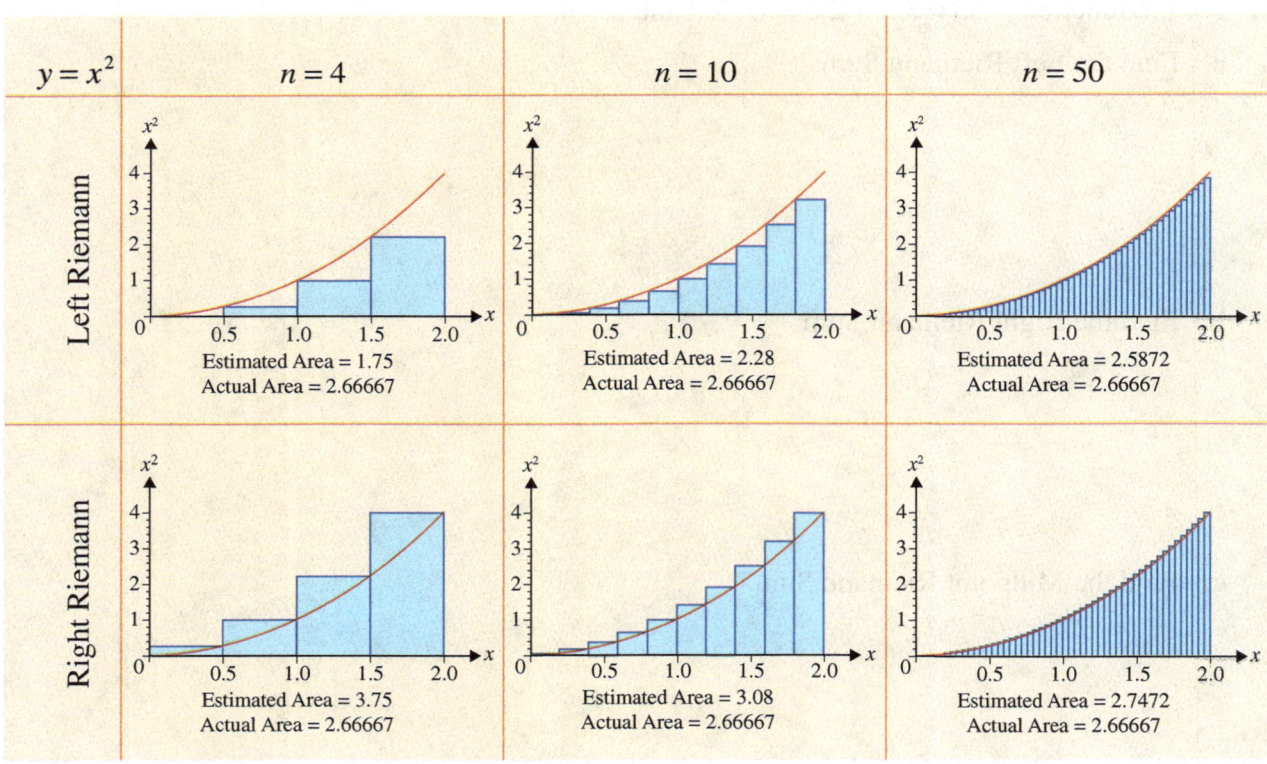

Section 5.1: Areas and Distances (Riemann Sums)

WeBWorK

4. The speed of a runner increased steadily during the first 3 seconds of a race. Her speed at half-second intervals is given in the table:

Time (s)	0	0.5	1	1.5	2	2.5	3
Velocity (ft/s)	0.7	3.5	5.3	8.2	12.2	13.2	17.8

   a. Find a lower estimate for the distance that she traveled during these 3 seconds.

   b. Find an upper estimate for the distance that she traveled during these 3 seconds.

The Definite Integral

Section 5.2

Before Class Video Examples

1. Approximate using the midpoint rule with $n = 3$ subintervals is used to approximate $\int_1^7 \ln x \, dx$. (Round your answer to four decimal places.)

2. Evaluate the integral $\int_{-2}^{4} (2x - 4) \, dx$ by sketching a graph and finding areas.

Section 5.2: The Definite Integral

3. If $\int_{-2}^{10} f(x)\,dx = 14.8$ and $\int_{5}^{10} f(x)\,dx = 18.3$, find $\int_{-2}^{5} f(x)\,dx$ using properties of the definite integral.

4. Using properties of integrals, how would evaluate the definite integral: $\int_{0}^{2} 8x^3 + 3x^2 \, dx =$

Algebra Review

1. **Limits**

   Example

   i. $\lim\limits_{n\to\infty} 5 + \dfrac{2}{n} =$

   ii. $\lim\limits_{n\to\infty} \dfrac{n^2 + 2n - 1}{2n^2 + 6} =$

2. **Expanding Polynomials**

   Example

   Expand and simplify:

   i. $\dfrac{x(x+1)(x+5)}{4} \cdot \dfrac{12}{x^2}$

Section 5.2: The Definite Integral

A. The Sum Operator

$$\sum_{i=1}^{n} x_i = x_1 + x_2 + \ldots + x_n$$

Examples

1. $\sum_{a=1}^{6} a = 1 + 2 + 3 + 4 + 5 + 6 = 21$

2. $\sum_{i=1}^{5} 3i = 3(1) + 3(2) + 3(3) + 3(4) + 3(5) = 45$

3. $\sum_{x=0}^{3} (2x+1) =$

4. $\sum_{i=1}^{n} a =$

Summation Rules and Properties

1. $\sum_{i=1}^{n} c = c \cdot n$	4. $\sum_{i=1}^{n} i = \dfrac{n \cdot (n+1)}{2}$
2. $\sum_{i=1}^{n} c \cdot a_i = c \cdot \sum_{i=1}^{n} a_i$	5. $\sum_{i=1}^{n} i^2 = \dfrac{n \cdot (n+1) \cdot (2n+1)}{6}$
3. $\sum_{i=1}^{n} a_i \pm b_i = \sum_{i=1}^{n} a_i \pm \sum_{i=1}^{n} b_i$	6. $\sum_{i=1}^{n} i^3 = \dfrac{n^2 \cdot (n+1)^2}{4}$

*Also Remember $\lim_{n \to \infty} \dfrac{c}{n} = 0$ and $\lim_{n \to \infty} \dfrac{c}{n^p} = 0$ for any constant c and any power $p \geq 1$.

Section 5.2: The Definite Integral

B. Definition of the Definite Integral

$$\int_a^b f(x)\,dx = \lim_{n\to\infty} \sum_{i=1}^n f(x_i) \cdot \Delta x \qquad \text{where } \Delta x = \frac{b-a}{n} \text{ and } x_i = a + i \cdot \Delta x$$

Example

5. Find the area under the curve of the function $f(x) = 3x^2 + 2$ on the interval $[0, 2]$.

The Exact area $= \displaystyle\int_0^2 3x^2 + 2\,dx = \lim_{n\to\infty} \sum_{i=1}^n (3x_i^2 + 2) \cdot \Delta x$

16   Section 5.2: The Definite Integral

WeBWorK

2. Find the Riemann sum for $f(x) = x^3, 1 \le x \le 11$

   a. If the partition points are 1, 3, 8, 11 and the sample points are 2, 5, 9.

   b. If the partition points are 1, 3, 8, 11 and the sample points are the midpoints.

7. The following area $\left(1+\dfrac{6}{n}\right)^2 \cdot \dfrac{6}{n} + \left(1+\dfrac{12}{n}\right)^2 \cdot \dfrac{6}{n} + \left(1+\dfrac{18}{n}\right)^2 \cdot \dfrac{6}{n} + \cdots \left(1+\dfrac{6n}{n}\right)^2 \cdot \dfrac{6}{n} +$ is a right Riemann sum for a certain definite integral $\int_0^b f(x)\,dx$ using a partition of the interval $[0, b]$ into n subintervals of equal length. Find the upper limit of integration b and the integrand function $f(x)$.

Section 5.2: The Definite Integral

9. Evaluate the integral $\int_{-4}^{4} 2+\sqrt{16-x^2}\, dx$ by interpreting in terms of areas.

14. Let $\int_{-10}^{-7} f(x)\, dx = 1$, $\int_{-10}^{-9} f(x)\, dx = 5$ and $\int_{-8}^{-7} f(x)\, dx = 9$,

   a. find $\int_{-9}^{-8} f(x)\, dx = \underline{}$ and

   b. $\int_{-8}^{-9} 2f(x) - 5\, dx = \underline{}$

Section 5.2: The Definite Integral

16. Given that $3 \leq f(x) \leq 5$ for $-5 \leq x \leq 7$, estimate the value of $\int_{-5}^{7} f(x)\,dx$.

    _____ $\leq \int_{-5}^{7} f(x)\,dx \leq$ _____

17. Use property 8 to estimate the value of the integral $\int_{3}^{13} \frac{7}{x}\,dx$

    (Property 8: If $m \leq f(x) \leq M$ for $a \leq x \leq b$, then $m(b-a) \leq \int_{a}^{b} f(x)\,dx \leq M(b-a)$)

    _____ $\leq \int_{3}^{13} \frac{7}{x}\,dx \leq$ _____

18. Express the following limit as a definite integral: $\lim_{n \to \infty} \sum_{i=1}^{n} \frac{i^6}{n^7} = \int_{a}^{b} f(x)\,dx$. Give a, b, and $f(x)$.

Evaluating Integrals

Section 5.3

Before Class Video Examples

1. Find $\int_2^3 9x^2\,dx$

2. Find $\int 4x^3 + \dfrac{5}{x} - \dfrac{8}{x^4} + 6\cos x + 7e^x - 4\sqrt{x} + 2\,dx$

3. Find $\int_{-2}^2 (x^2+3)(2x-1)\,dx$

Section 5.3: Evaluating Integrals

4. Find $\int \dfrac{9x^2 + 7}{\sqrt{x}} dx$

Algebra Review

1. Exponents

Exponential Notation

$b^n = \underbrace{b \cdot b \cdot b \cdot \ldots \cdot b}_{n \text{ times}}$, Example: $3^4 = 3 \cdot 3 \cdot 3 \cdot 3 = 81$

Common Rules of Exponents

- $x^a \cdot x^b = x^{(a+b)}$ (Product Rule)
- $\dfrac{x^a}{x^b} = x^{(a-b)}$ (Quotient Rule)
- $x^0 = 1$ (Zero-Exponent Rule)
- $\dfrac{1}{x^b} = x^{(-b)}$ or $\dfrac{1}{x^{-b}} = x^b$ (Negative Exponent Rule)
- $(x^a)^b = x^{a \cdot b}$ (Power Rule)
- $(x \cdot y)^a = x^a \cdot y^a$ (Products to Power)
- $\left(\dfrac{x}{y}\right)^a = \dfrac{x^a}{y^a}$ (Quotients to Power)

Example

i. $(x^3)^5 =$

ii. $\dfrac{2}{x^3} =$

Section 5.3: Evaluating Integrals

iii. $(2x^4)(7x^3) =$

iv. $e^{x+1} =$

v. $3^{2x} =$

2. Radicals and Rational Exponents

Common Rules of Radicals

- $\sqrt{x \cdot y} = \sqrt{x} \cdot \sqrt{y}$ (Product Rule)
- $\sqrt{\dfrac{x}{y}} = \dfrac{\sqrt{x}}{\sqrt{y}}$, $y \neq 0$ (Quotient Rule)
- $x^{\frac{m}{n}} = \sqrt[n]{x^m} = \left(\sqrt[n]{x}\right)^m$
- $\sqrt[n]{x} = y \Leftrightarrow y^n = x$ (n^{th} root)
- $x^{-\frac{1}{n}} = \dfrac{1}{x^{\frac{1}{n}}} = \dfrac{1}{\sqrt[n]{x}}$ (Rational Exponents)

Example

vi. $\sqrt{x} =$

vii. $\sqrt[7]{x^5} =$

viii. $\dfrac{3x^2 + 7x}{\sqrt{x}} =$

Section 5.3: Evaluating Integrals

3. Polynomial Products

Expanding Binomials

FOIL:
$$(a+b)(c+d) = ac + ad + bc + bd$$
First — Outer — Inner — Last

Example

ix. $(2x+3)(x-1) =$

A. Notation

$$\int f(x)\,dx = F(x) + C \quad \text{or} \quad \int f'(x)\,dx = f(x) + C$$

B. Fundamental Theorem of Calculus

$$\int_a^b f(x)\,dx = F(b) - F(a)$$

Section 5.3: Evaluating Integrals

Definition: Definite integral $\int_a^b f(x)\,dx = F(b) - F(a)$ and an

Indefinite integral $\int f(x)\,dx = F(x) + C$

Table of Indefinite Integrals

1. $\int x^n\,dx = \dfrac{x^{n+1}}{n+1} + C \ (n \neq -1)$

2. $\int \dfrac{1}{x}\,dx = \ln|x| + C$

3. $\int e^x\,dx = e^x + C$

4. $\int a^x\,dx = \dfrac{a^x}{\ln a} + C$

5. $\int \sin x\,dx = -\cos x + C$

6. $\int \cos x\,dx = \sin x + C$

7. $\int \sec^2 x\,dx = \tan x + C$

8. $\int \csc^2 x\,dx = -\cot x + C$

9. $\int \sec x \cdot \tan x\,dx = \sec x + C$

10. $\int \csc x \cdot \cot x\,dx = -\csc x + C$

11. $\int \dfrac{1}{x^2+1}\,dx = \tan^{-1} x + C$

12. $\int \dfrac{1}{\sqrt{1-x^2}}\,dx = \sin^{-1} x + C$

Section 5.3: Evaluating Integrals

Properties of the Integral

1. $\int_a^b k\, dx = k(b-a)$ for a constant k

2. $\int_a^b [f(x) \pm g(x)]\, dx = \int_a^b f(x)\, dx \pm \int_a^b g(x)\, dx$

3. $\int_a^b k \cdot f(x)\, dx = k \cdot \int_a^b f(x)\, dx$

4. $\int_a^b f(x)\, dx = -\int_b^a f(x)\, dx$

5. $\int_a^b f(x)\, dx = 0$ if $a = b$

6. $\int_a^b f(x)\, dx + \int_b^c f(x)\, dx = \int_a^c f(x)\, dx$ also

   $\int_a^c f(x)\, dx - \int_b^c f(x)\, dx = \int_a^b f(x)\, dx$

7. $\int_a^b f(x)\, dx \geq 0$ if $f(x) \geq 0$ and $a \leq x \leq b$

8. $\int_a^b f(x)\, dx \geq \int_a^b g(x)\, dx$ if $f(x) \geq g(x)$ and $a \leq x \leq b$

Examples

1. $\int 3x^2 + 2\, dx$

2. $\int 3x^4 - \dfrac{7}{x} + 4e^x - 5\sin x + 3\, dx$

Section 5.3: Evaluating Integrals

3. $\int 3x^2 + \dfrac{1}{x} - \dfrac{6}{x^2} + \sqrt{x} + 6\sec x \tan x + 2 \, dx$

4. $\int \dfrac{3t^2 + 7t + 5}{\sqrt{t}} \, dt$

5. $\int_0^2 3x^2 + 2 \, dx$

26 Section 5.3: Evaluating Integrals

C. "Negative" Area/Area under the x-Axis

The integral calculates the area under the graph of a function $f(x)$. More specifically, the integral of a function $f(x)$ is equal to the total area above the x-axis plus the negative value of the total below the x-axis.

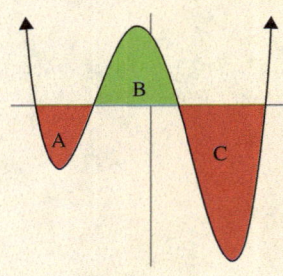

$$\int_{-4}^{4} f(x)\,dx = (-A) + B + (-C)$$

Examples

7. (Type 1: Find the integral using the area.) Find $\int_{0}^{3} f(x)\,dx$ using geometry.

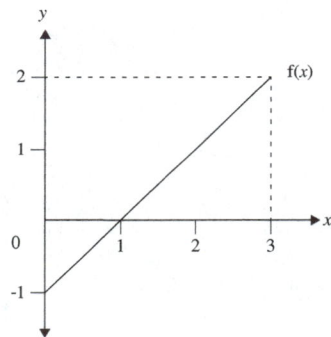

8. (Type 2: Find the area using the integral.) Find the total shaded area on the graph related to the function $f(x) = \sin x$

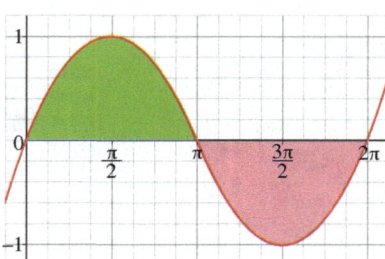

Section 5.3: Evaluating Integrals   27

WeBWorK

15. $\int 9e^{u+1}\, du =$

19. $\int \dfrac{5\sin t}{1-\sin^2 t}\, dt =$

20. The velocity of a function is $v(t) = t^2 - 6t + 8$ for a particle moving along a line. Find the displacement and the distance traveled by the particle during the time interval [0, 5].

Hint: Displacement $= \int_0^5 v(t)\, dt$ and Distance traveled $= \int_0^5 |v(t)|\, dt$

Note that $t^2 - 6t + 8 = (t-2)(t-4)$. Find the interval(s) where $v(t) \leq 0$ and $v(t) \geq 0$

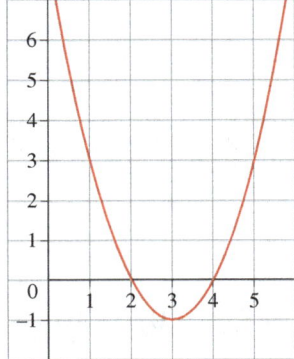

Section 5.3: Evaluating Integrals

21. The acceleration function for a particle moving along a line is $a(t) = 2t + 3$.
The initial velocity is $v(0) = -10$. Then

a. The velocity at time t, $v(t) =$

b. The distance traveled during the time interval [0, 4] is equal to =

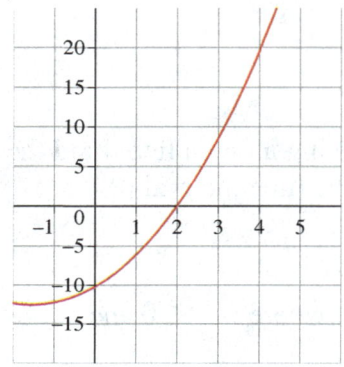

23. Suppose h is a function such that $h(1) = -8$, $h'(1) = 2$, $h''(1) = 7$, $h(10) = 4$, $h'(10) = -10$, $h''(10) = 17$ and h'' is continuous everywhere. Then $\displaystyle\int_1^{10} h''(u)\, du =$

The Fundamental Theorem of Calculus

Section 5.4

Before Class Video Examples

1. If $f(x) = \int_{14}^{x} 5e^{3t^2+10} dt$, then $f'(x) =$

2. If $f(x) = \int_{x}^{102} \sin(e^{2t}) + 17 dt$, then $f'(x) =$

3. If $f(x) = \int_{98}^{4x^3} 5\ln(t^3 + 15) dt$, then $f'(x) =$

4. Find the Average Value of the function $f(x) = \dfrac{15}{x}$ on the interval $[1, 12]$

29

Section 5.4: The Fundamental Theorem of Calculus

Algebra/Calc I Review

1. **Chain Rule**

$$\frac{d}{dx}\bigl(f(g(x))\bigr) = f'(g(x)) \cdot g'(x)$$

Example
Give $f'(x)$

i. $f(x) = e^{x^2}$

ii. $f(x) = (\ln x)^4$

2. **Solving Equations**

Example
Solve for x

iii. $z^2 + 6x^{-4} = 3 + z^2$

Section 5.4: The Fundamental Theorem of Calculus

A. Fundamental Theorem of Calculus—Part I

If f is a continuous function on the interval $[a,b]$, and $g(x) = \int_a^x f(t)\, dt$ where $a \leq x \leq b$, then $g'(x) = f(x)$

Reminder: Leibniz Notation $\dfrac{d}{dx}\left[\int_a^x f(t)\, dt\right] = f(x)$

Variations on the rule:

i. $\dfrac{d}{dx}\left[\int_a^x f(t)\, dt\right] = f(x)$

ii. $\dfrac{d}{dx}\left[\int_x^a f(t)\, dt\right] = -f(x)$

iii. $\dfrac{d}{dx}\left[\int_a^{g(x)} f(t)\, dt\right] = f(g(x)) \cdot g'(x)$

iv. $\dfrac{d}{dx}\left[\int_{h(x)}^{g(x)} f(t)\, dt\right] = f(g(x)) \cdot g'(x) - f(h(x)) \cdot h'(x)$

Examples

1. $\dfrac{d}{dx}\left[\int_3^x \cos t\, dt\right] =$

2. $\dfrac{d}{dx}\left[\int_{107}^x 1 + \sqrt{9 - t^2}\, dt\right] =$

3. $\dfrac{d}{dy}\left[\int_{107}^y \dfrac{x^3 + x^2 - 4x}{\cos x + 12}\, dx\right] =$

Section 5.4: The Fundamental Theorem of Calculus

4. $\dfrac{d}{dx}\left[\displaystyle\int_x^4 \tan t + \sec^2 t + e^{\cos t}\, dt\right] =$

5. $\dfrac{d}{dx}\left[\displaystyle\int_5^{x^2} \sqrt{t^2 + 14t}\, dt\right] =$

6. $\dfrac{d}{dx}\left[\displaystyle\int_{x^2}^{\sin x} \sqrt{t+5}\, dt\right] =$

7. Find the derivative of the function $f(x) = \displaystyle\int_{e^x}^{2x} \cos(t^2)\, dt$. (And do not forget to label!)

Section 5.4: The Fundamental Theorem of Calculus

B. Average Value of a Function

$$f_{ave} = \frac{1}{b-a} \cdot \int_a^b f(x)\,dx$$

Example

8. Find the average value of the function $f(x) = x^2 - 1$ on the interval $[1, 3]$

WeBWorK

1. Let $f(x) = \begin{cases} 0 & \text{if } x < -5 \\ 2 & \text{if } -5 \leq x < 2 \\ 4 - x & \text{if } 2 \leq x < 6 \\ -2 & \text{if } x \geq 6 \end{cases}$

and $g(x) = \int_{-5}^{x} f(t)\,dt$

Determine the following:

a. $g(-10) =$ 　　　　　　　　　　　$g(-3) =$

　　$g(2) =$ 　　　　　　　　　　　　$g(6) =$

　　$g(8) =$

b. $g(x)$ is increasing on the interval (A, B) where A = _____ and B = _____

c. The absolute maximum of $g(x) =$ _____ and occurs when $x =$ _____

Section 5.4: The Fundamental Theorem of Calculus

10. a. Find the average value of $f(x) = 25 - x^2$ on the interval $[0, 4]$.

b. Find a value c in the interval $[0, 4]$ such that $f(c)$ is equal to the average value.

11. Consider the function $f(x) = \begin{cases} x & \text{if } x < 1 \\ \dfrac{1}{x} & \text{if } x \geq 1 \end{cases}$

a. Evaluate the definite integral $\displaystyle\int_{-4}^{6} f(x)\, dx$

b. Evaluate the average value of f on the interval $[-4, 6]$

Section 5.4: The Fundamental Theorem of Calculus

14. Find a function f and a positive number a such that $2 + \int_a^x \frac{f(t)}{t^5} dt = 3x^{-2}$, $x > 0$

    a. $f(x) =$

    b. $a =$

Integration by U-Substitution

Section 5.5

Before Class Video Examples

1. Find $\int 5e^{3x}\,dx$

2. Find $\int x^3 \sin(2x^4)\,dx$

3. Find $\int \cos^3 x \sin x\,dx$

4. Find $\int \dfrac{x+3}{x^2+6x+15}\,dx$

5. Find $\int_0^2 x^2 e^{x^3-6}\,dx$

Section 5.5: Integration by U-Substitution

Algebra Review

1. **Composite functions**

Notation	$(f \circ g)(x) = f(g(x))$

   Example

   i. For the functions $f(x) = \sqrt{x}$ and $g(x) = x+2$ find $(f \circ g)(x) =$

   ii. For the function $(f \circ g)(x) = \sqrt{x+2}$, give $f(x)$ and $g(x)$

   iii. For the function $(f \circ g \circ h)(x) = \sin(e^{2x})$, give $f(x)$, $g(x)$, and $h(x)$

2. **Trigonometric Identities**

   Reciprocal Identities
   $$\cot x = \frac{1}{\tan x} = \frac{\cos x}{\sin x}$$
   $$\csc x = \frac{1}{\sin x}$$
   $$\sec x = \frac{1}{\cos x}$$

   Pythagorean Identities
   $$\sin^2 x + \cos^2 x = 1$$
   $$\sin^2 x = 1 - \cos^2 x$$
   $$\cos^2 x = 1 - \sin^2 x$$
   $$1 + \tan^2 x = \sec^2 x$$
   $$1 + \cot^2 x = \csc^2 x$$

   Sum And Difference Identities
   $$\sin(x \pm y) = \sin x \cos y \pm \cos x \sin y$$
   $$\cos(x \pm y) = \cos x \cos y \mp \sin x \sin y$$
   $$\tan(x \pm y) = \frac{\tan x \pm \tan y}{1 \mp \tan x \tan y}$$

   Double-Angle Identities
   $$\sin 2x = 2 \sin x \cos x = \frac{2 \tan x}{1 + \tan^2 x}$$
   $$\cos 2x = \cos^2 x - \sin^2 x$$
   $$= 2 \cos^2 x - 1$$
   $$= 1 - 2 \sin^2 x = \frac{1 - \tan^2 x}{1 + \tan^2 x}$$
   $$\tan 2x = \frac{2 \tan x}{1 - \tan^2 x}$$
   $$\cot 2x = \frac{\cot^2 x - 1}{2 \cot x}$$

   Half-Angle Identities
   $$\sin \frac{x}{2} = \pm \sqrt{\frac{1 - \cos x}{2}}$$
   $$\cos \frac{x}{2} = \pm \sqrt{\frac{1 + \cos x}{2}}$$
   $$\tan \frac{x}{2} = \pm \sqrt{\frac{1 - \cos x}{1 + \cos x}} = \frac{\sin x}{1 + \cos x} = \frac{1 - \cos x}{\sin x}$$

   Example

   iv. $1 - \cos^2(\theta) =$

A. U-Substitution

We will use integration by substitution when we are trying to find an integral of product and quotient of functions or of composite functions. We will try to find a component of the function (which we will set equal to u), such that if we find the derivative of that component, the result (multiplied by a constant) is something that is part of the original function.

Example: $\int (2x+1)(x^2+x-5)^5 \, dx$

Let's first identify the two components in this function:

- Component 1 is $(2x+1)$
- Component 2 is (x^2+x-5)

If we take a derivative of component 2 (x^2+x-5), we will get component 1 $(2x+1)$. For that reason, we will let component 2 (x^2+x-5) be u in our problem. So,

$$u = x^2 + x - 5$$

Next, we will find the derivative of u:

$$\frac{du}{dx} = 2x+1 \quad \Rightarrow \quad du = (2x+1)\,dx$$

We are now ready to make some substitutions. In the integral $\int (2x+1)(x^2+x-5)^5 \, dx$ we will replace

- $u = x^2 + x - 5$
- $du = (2x+1)\,dx$

$$\Rightarrow \int (2x+1)(x^2+x-5)^5 \, dx = \int (x^2+x-5)^5 (2x+1)\,dx = \int (u)^5 \, du$$

We will integrate $\int (u)^5 \, du = \dfrac{u^6}{6}$

And then back substitute $u = x^2 + x - 5$ into the result to get $\dfrac{(x^2+x-5)^6}{6}$

Finally, we can conclude that $\int (2x+1)(x^2+x-5)^5 \, dx = \dfrac{(x^2+x-5)^6}{6} + C$

Section 5.5: Integration by U-Substitution

Examples

1. $\int (x+1)(x^2+2x+5)^8 \, dx$

2. $\int x \cdot \sqrt{x^2+4} \, dx$

3. $\int x^2 \cdot e^{x^3} \, dx$

4. $\int \dfrac{x+2}{x^2+4x+6}\,dx$

5. $\int \sin^2 x \cos x \, dx$

6. $\int \dfrac{\ln x}{x}\,dx$

Section 5.5: Integration by U-Substitution

7. $\int \tan \theta \, d\theta$

8. $\int \dfrac{1+x}{1+x^2} \, dx$

9. $\int \sec^3 x \cdot \tan x \, dx$

Section 5.5: Integration by U-Substitution

10. $\int (x+1)\sin(x^2+2x+15)\,dx$

11. $\int x \cdot \sqrt{x+4}\,dx$

Section 5.5: Integration by U-Substitution

B. Common Forms

$$\int e^{kx}\, dx = \frac{1}{k}e^{kx} + C$$

$$\int \sin(kx)\, dx = -\frac{1}{k}\cos(kx) + C$$

$$\int \cos(kx)\, dx = \frac{1}{k}\sin(kx) + C$$

Examples

12. $\int e^{5x}\, dx$

13. $\int \sin(\pi x)\, dx$

C. Change of Bounds

Example

14. By letting $u = x + 2$, the integral $\int_{1}^{3}(x+2)^3\, dx$ can be expressed as the integral $\int_{A}^{B} u^3\, du$. Find the values of A and B.

Section 5.5: Integration by U-Substitution

D. Calculator

Example

15. Use your calculator to evaluate $\int_0^4 \frac{x}{\sqrt{1+2x}}\, dx$

TI 83/84 (Older) — MATH → Choose 9: **fnInt (** → Enter the **fxn, x, lowerbound , upperbound)** — ENTER

TI 84 (Newer) — MATH → Choose 9: → Enter information into palette — ENTER

WeBWorK

14. $\int \frac{e^{2x}+6}{e^{2x}}\, dx$

17. $\int \frac{6x}{1+x^4}\, dx$

Section 5.5: Integration by U-Substitution

19. $\int 12\sec^2\left(\dfrac{t}{4}\right) dt$

23. $\int \dfrac{x}{\sqrt{x+9}}\, dx$

24. $\displaystyle\int_{-\pi/4}^{\pi/4} x^6 \tan^7 x\, dx$

28. If f is continuous and $\displaystyle\int_0^4 f(x)\, dx = 13$, then $\displaystyle\int_0^2 f(2x)\, dx =$

Integration by Parts

Section 6.1

Before Class Video Examples

1. Find $\int 3xe^{6x}\,dx$

2. Find $\int_0^1 3xe^{6x}\,dx$

3. Find $\int 2x\cos(5x)\,dx$

4. Find $\int \dfrac{\ln x}{x^3}\,dx$

Section 6.1: Integration by Parts

Algebra Review

1. **Families of Functions**

Algebraic	Example: 2, x, x^2+5x+1, $\sqrt{x}$, $\dfrac{1}{x^4}$
Trigonometric	Example: $\sin x$, $\tan(\theta)$
Inverse Trigonometric	Example: $\cos^{-1} x$, $\arctan(\theta)$
Logarithmic	Example: $\ln x$, $\log_5(x)$
Exponential	Example: e^x, 7^{2x}

Example

Identify each of the following functions by family

i. $f(x) = \arcsin x$

ii. $g(x) = 1$

iii. $h(\beta) = \sec \beta$

iv. $m(x) = e^{x+1}$

v. $f(t) = \ln(t)$

vi. $m(x) = \dfrac{1}{\sqrt{x}}$

Section 6.1: Integration by Parts 49

A. Formula (and How to Use It)

$$\int u \, dv = u \cdot v - \int v \, du$$

To use the above formula

1. Look at the product that is given, and decide what u will be and what dv will be. (We will always put dx in our function with whatever we declare dv to be.)
2. Find du by taking the derivative of u
3. Find v by integrating dv
4. Plug into the formula and simplify
5. $\int v \, du$ should be something that you can easily integrate. If not, you may need to switch your choice of u and dv.

Example: $\int x \sin x \, dx$

We will let $\int [x][\sin x \, dx] = \int [u][dv]$

So $\quad u = x \quad$ and $\quad dv = \sin x \, dx$

$\Rightarrow \quad \dfrac{du}{dx} = 1 \quad\quad\quad\quad \Rightarrow \quad$ Divide each side by dx

$\Rightarrow \quad$ Multiply each side by $dx \quad\quad \Rightarrow \quad \dfrac{dv}{dx} = \sin x$

$\Rightarrow \quad du = dx \quad\quad\quad\quad\quad\quad \Rightarrow \quad v = \int \sin x \, dx$

$\quad\quad\quad\quad\quad\quad\quad\quad\quad\quad\quad\quad\quad \Rightarrow \quad v = -\cos x$

We may now use the formula $\int u \, dv = u \cdot v - \int v \, du$

$$\int x \cdot \sin x \, dx = x \cdot (-\cos x) - \int (-\cos x) \, dx$$
$$= -x \cdot \cos x + \int \cos x \, dx$$
$$= -x \cdot \cos x + \sin x + C$$

Section 6.1: Integration by Parts

Examples

1. $\int 3x \cdot e^{2x} \, dx$

2. $\int x^2 \cdot e^x \, dx$

3. $\int x^3 \cdot \ln x \, dx$

Section 6.1: Integration by Parts

4. $\int \ln x \, dx$

5. $\int \dfrac{\ln x}{x^6} \, dx$

Section 6.1: Integration by Parts

6. $\int e^{3x} \cdot \sin(2x)\, dx$

B. The LIATE Principle for Integration by Parts

A common question when using integrations by parts is: "What part of the equation should I let u be equal to and dv be equal to?" Although there is no "set in stone" answer, we can use the LIATE principle as a guideline. Recall that

$$\int u \, dv = u \cdot v - \int v \, du$$

This rule of thumb is for choosing the function that is to be u when using integration by parts.

Logarithmic functions (e.g., $\ln x$)

Inverse trigonometric functions (e.g., $\sin^{-1} x$)

Algebraic functions (e.g., $5x^3 + 4x^2 - x$ or $5x^3 + 4x^2 - x$)

Trigonometric functions (e.g., $\cos x$)

Exponential functions (e.g., e^x or 8^{2x})

The *higher* a type of function appears on this list, the more likely it should serve as u in the integration by parts formula. Conversely, the lower a type of function appears on this list, the more likely it should serve as v.

7. $\int \tan^{-1} x \, dx$

54 Section 6.1: Integration by Parts

WeBWorK

10. Use a combination of substitution and parts to evaluate the integral $\int \sin(3\sqrt{x})\,dx$.

Step 1: Substitution, let $w = 3\sqrt{x}$. Note: We use w here for the substitution instead of the more common variable u, since it is convenient for us to reserve u and v for the upcoming integration by parts.

It follows that $dw = \dfrac{3}{2\sqrt{x}}dx$ and $dx = \dfrac{2\sqrt{x}}{3}dw$. To successfully continue with substitution, it is now necessary to rewrite dx strictly in terms of w. Thus $dx = f(w)\,dw$, where

$f(w) = $ _____

Complete the substitution, to get $\int \sin(3\sqrt{x})\,dx = \int g(w)\,dw$ where $g(w) = $ _____

Step 2: Use integration by parts to integrate $\int g(w)\,dw$. Let $u = $ _____ and $dv = \sin(w)\,dw$.

This gives (as a function of w), $\int g(w)\,dw = $ _____ $+C$.

Step 3: Substitute $3\sqrt{x}$ for w, to get $\int \sin(3\sqrt{x})\,dx = $

_____ $+C$.

15. Suppose that $f(5) = 2$, $f(7) = 6$, $f'(5) = 9$, $f'(7) = 7$, and f'' is continuous.

Find the value of the definite integral: $\displaystyle\int_5^7 x \cdot f''(x)\, dx = $ _____

Partial Fractions and Additional Techniques of Integration

Section 6.3

Before Class Video Examples

1. Partial Fractions—Type 1: $\int \dfrac{17x-2}{x^2-x-12}dx$

   Step 1. Notice U-Substitution does not work and the Denominator Factors!

   Step 2. Factor the denominator and rewrite the fraction over the partial fraction terms.

   Step 3. Solve for the constants and plug back into the function.

   Step 4. Integrate the smaller fraction problem.

Section 6.3: Partial Fractions and Additional Techniques of Integration

2. Partial Fractions—Type 2: $\int \dfrac{2x^2 - 2x - 9}{x - 5} dx$

   Degree of the numerator is greater that the degree of the denominator. First do Long Division.

3. Arctangent Formula: $\int \dfrac{9}{x^2 + 16} dx$

Section 6.3: Partial Fractions and Additional Techniques of Integration

4. Combination of Long Division/Partial Fractions/Arctangent Formula: $\int \dfrac{3x^2 + x - 5}{x^2 + 9} dx$

Section 6.3: Partial Fractions and Additional Techniques of Integration

Algebra Review

1. **Common Denominators**

$$\frac{A}{B} + \frac{C}{D} = \frac{AD + BC}{BD}$$

Example

i. $\dfrac{2}{x+1} + \dfrac{5}{x-4} =$

2. **Factoring Quadratics**

$$AC + BC + AD + BD = (A+B)(C+D)$$

Example

ii. $x^2 + 2x - 15$

iii. $6x^2 + 8x + 2$

Section 6.3: Partial Fractions and Additional Techniques of Integration

3. Prime Functions

> Discriminant: For $ax^2 + bx + c$, the discriminant is $b^2 - 4ac$
>
> $ax^2 + bx + c$ is prime (i.e., cannot factor) if the discriminant $b^2 - 4ac < 0$

Example
Determine whether the function has factors (i.e., the function has real zeros)

iv. $x^2 + 2x - 15$

v. $2x^2 + 2x + 8$

4. Generic Expressions

Degree 0 (i.e., constant)	C or K	e.g., $2, 0, \pi, \frac{1}{5}$
Degree 1	$Ax + B$, where $A \neq 0$	e.g., $x + 1, -\frac{1}{2}x$
Degree 2	$Ax^2 + Bx + C$, where $A \neq 0$	e.g., $x^2 + 5x + 4, 5x^2 + 5$
Degree n	$a_n x^n + a_{n-1} x^{n-1} + \ldots + a_1 x^1 + C$, where $a_n \neq 0$	e.g., $2x^8 - 10x^4, 5x^2 + 5$

Section 6.3: Partial Fractions and Additional Techniques of Integration

5. Proper Fractions

> A rational expression qualifies as a proper fraction if the denominator has a degree that is strictly greater than the degree of the numerator.

Example

Provide a possible numerator for each of these denominators so that the expression will be a proper fraction

vi. $\dfrac{}{x^2 + 2x - 15}$

vii. $\dfrac{}{7x + 1}$

viii. $\dfrac{}{2x^3 - 12x + 3}$

Section 6.3: Partial Fractions and Additional Techniques of Integration

A. Partial Fractions

Examples

1. $\displaystyle\int \frac{2x+1}{x^2 - 3x - 10}\,dx$

Section 6.3: Partial Fractions and Additional Techniques of Integration

2. $\int \dfrac{x-1}{(x+1)(x^2+x+1)}\,dx$

Section 6.3: Partial Fractions and Additional Techniques of Integration

Power Rule for Partial Fractions

Consider the Expression: $\dfrac{Ax+C}{(x-1)^2}$ We can expand this expression so that

$$\dfrac{Ax+C}{(x-1)^2} = \dfrac{(Ax-A)+(A+C)}{(x-1)^2} = \dfrac{(Ax-A)}{(x-1)^2}+\dfrac{(A+C)}{(x-1)^2} = \dfrac{A(x-1)}{(x-1)^2}+\dfrac{(A+C)}{(x-1)^2}$$

In the first term, the $(x-1)$ terms can cancel out, in the second term, we can replace $A+C = B$, as A, B, and C all represent random constants. This gives

$$\dfrac{A(x-1)}{(x-1)^2}+\dfrac{(A+C)}{(x-1)^2} = \dfrac{A}{(x-1)}+\dfrac{B}{(x-1)^2}$$

Example

3. Give a generic expansion for the following:

a. $\dfrac{3x^2+5}{(x-2)^2(x+1)}$

b. $\dfrac{x^4}{(x+1)^3(x-5)^2}$

c. $\dfrac{4x^2-2}{(x+8)^2(x^2+1)}$

d. $\dfrac{5x^3-1}{(x^2+2x+1)^2}$

e. $\dfrac{x^2+2}{x^2(x-2)}$

Section 6.3: Partial Fractions and Additional Techniques of Integration

4. $\int \dfrac{3x^2 + 5}{(x-2)^2 (x+1)}\, dx$

Section 6.3: Partial Fractions and Additional Techniques of Integration

B. Other Techniques: Formulae

$$\int \frac{1}{u^2 + a^2} du = \frac{1}{a} \tan^{-1}\left(\frac{u}{a}\right) + C$$

Examples

5. $\int \dfrac{1}{x^2 + 9} dx$

6. $\int \dfrac{1}{4x^2 + 9} dx$

Section 6.3: Partial Fractions and Additional Techniques of Integration

C. Long Division

Perform long division when degree of numerator ≥ degree of denominator.

Examples

7. Find the quotient $\dfrac{x^4 - 5x^2 + x - 7}{x - 2}$

8. $\displaystyle\int \dfrac{x^2}{x+4}\, dx$

Section 6.3: Partial Fractions and Additional Techniques of Integration

9. $\displaystyle\int \frac{x^3 - 4x - 10}{x^2 - x - 6}\, dx$

Section 6.3: Partial Fractions and Additional Techniques of Integration

D. Completing the Square

> If a quadratic expression $x^2 + bx + c$ has the quality $c = \left(\dfrac{b}{2}\right)^2$ then it will factor into a perfect square: $x^2 + bx + c = \left(x + \dfrac{b}{2}\right)^2$

Examples

10. Complete the square: $x^2 + 6x + 13 =$

11. $\displaystyle\int \dfrac{1}{x^2 + 2x + 10}\, dx$

Section 6.3: Partial Fractions and Additional Techniques of Integration

12. $\displaystyle\int \frac{21}{2x^2 + 6x + 15}\,dx$

Section 6.3: Partial Fractions and Additional Techniques of Integration

WeBWorK

15. Substitute $u = \sqrt{x}$ to express the integrand as a rational function and then evaluate $\int \dfrac{2\sqrt{x}}{x+9}\,dx$

16. Substitute $u = e^x$ to express the integrand as a rational function and then evaluate

$$\int \frac{-16e^x - 40}{e^{2x} + 6e^x + 8} \, dx$$

Integration with Tables

Section 6.4

Before Class Video Examples

1. From the Tables of Integrals in the back of your book we have:

$$\int \sqrt{a^2 - u^2}\, du = \frac{u}{2}\sqrt{a^2 - u^2} + \frac{a^2}{2}\sin^{-1}\left(\frac{u}{2}\right) + C$$

Use it to find the following: $\int \sqrt{25 - 9x^2}\, dx$

2. From the Tables of Integrals in the back of your book we have:

$$\int \sin^2 u\, du = \frac{1}{2}u - \frac{1}{4}\sin(2u) + C$$

Use it to find the following: $\int \sin^2(5x + 1)\, dx$

Section 6.4: Integration with Tables

3. From the Tables of Integrals in the back of your book we have:

$$\int \frac{u^2}{\sqrt{a^2-u^2}}\,du - \frac{u}{2}\sqrt{a^2-u^2} + \frac{a^2}{2}\sin^{-1}\left(\frac{u}{a}\right) + C$$

Use it to find the following: $\int \frac{5x^2}{\sqrt{17-9x^2}}\,dx$

4. From the Tables of Integrals in the back of your book we have:

$$\int \frac{\sqrt{a^2-u^2}}{u}\,du = \sqrt{a^2-u^2} - a\ln\left|\frac{a+\sqrt{a^2-u^2}}{u}\right| + C$$

Use it to find the following: $\int \sqrt{25-e^{2x}}\,dx$

Section 6.4: Integration with Tables   77

Algebra Review

1. **Expressions as Square Values**

$$x = \left(\sqrt{x}\right)^2$$

Example

Give the following expressions as a single value squared

i. $4x^2$

ii. 25

iii. 5

iv. $8x^2$

v. y^4

vi. e^{2x}

Give the missing value

vii. $(4x)^2 = a \cdot 8x^2$

viii. $(a \cdot 3y)^2 = 81y^2$

Section 6.4: Integration with Tables

A. Tables

TABLE OF INTEGRALS

BASIC FORMS

1. $\int u\, dv = uv - \int v\, du$

2. $\int u^n\, du = \dfrac{u^{n+1}}{n+1} + C,\ n \neq -1$

3. $\int \dfrac{du}{u} = \ln|u| + C$

4. $\int e^u\, du = e^u + C$

5. $\int a^u\, du = \dfrac{a^u}{\ln a} + C$

6. $\int \sin u\, du = -\cos u + C$

7. $\int \cos u\, du = \sin u + C$

8. $\int \sec^2 u\, du = \tan u + C$

9. $\int \csc^2 u\, du = -\cot u + C$

10. $\int \sec u \tan u\, du = \sec u + C$

11. $\int \csc u \cot u\, du = -\csc u + C$

12. $\int \tan u\, du = \ln|\sec u| + C$

13. $\int \cot u\, du = \ln|\sin u| + C$

14. $\int \sec u\, du = \ln|\sec u + \tan u| + C$

15. $\int \csc u\, du = \ln|\csc u - \cot u| + C$

16. $\int \dfrac{du}{\sqrt{a^2 - u^2}} = \sin^{-1}\dfrac{u}{a} + C$

17. $\int \dfrac{du}{a^2 + u^2} = \dfrac{1}{a}\tan^{-1}\dfrac{u}{a} + C$

18. $\int \dfrac{du}{u\sqrt{u^2 - a^2}} = \dfrac{1}{a}\sec^{-1}\dfrac{u}{a} + C$

19. $\int \dfrac{du}{a^2 - u^2} = \dfrac{1}{2a}\ln\left|\dfrac{u+a}{u-a}\right| + C$

20. $\int \dfrac{du}{u^2 - a^2} = \dfrac{1}{2a}\ln\left|\dfrac{u-a}{u+a}\right| + C$

TABLE OF INTEGRALS

FORMS INVOLVING $\sqrt{a^2+u^2}, a>0$

21. $\int \sqrt{a^2+u^2}\, du = \dfrac{u}{2}\sqrt{a^2+u^2} + \dfrac{a^2}{2}\ln\left(u+\sqrt{a^2+u^2}\right) + C$

22. $\int u^2\sqrt{a^2+u^2}\, du = \dfrac{u}{8}(a^2+2u^2)\sqrt{a^2+u^2} - \dfrac{a^4}{8}\ln\left(u+\sqrt{a^2+u^2}\right) + C$

23. $\int \dfrac{\sqrt{a^2+u^2}}{u}\, du = \sqrt{a^2+u^2} - a\ln\left|\dfrac{a+\sqrt{a^2+u^2}}{u}\right| + C$

24. $\int \dfrac{\sqrt{a^2+u^2}}{u^2}\, du = -\dfrac{\sqrt{a^2+u^2}}{u} + \ln\left(u+\sqrt{a^2+u^2}\right) + C$

25. $\int \dfrac{du}{\sqrt{a^2+u^2}} = \ln\left(u+\sqrt{a^2+u^2}\right) + C$

26. $\int \dfrac{u^2\, du}{\sqrt{a^2+u^2}} = \dfrac{u}{2}\sqrt{a^2+u^2} - \dfrac{a^2}{2}\ln\left(u+\sqrt{a^2+u^2}\right) + C$

27. $\int \dfrac{du}{u\sqrt{a^2+u^2}} = -\dfrac{1}{a}\ln\left|\dfrac{\sqrt{a^2+u^2}+a}{u}\right| + C$

28. $\int \dfrac{du}{u^2\sqrt{a^2+u^2}} = -\dfrac{\sqrt{a^2+u^2}}{a^2 u} + C$

29. $\int \dfrac{du}{(a^2+u^2)^{3/2}} = \dfrac{u}{a^2\sqrt{a^2+u^2}} + C$

FORMS INVOLVING $\sqrt{a^2-u^2}, a>0$

30. $\int \sqrt{a^2-u^2}\, du = \dfrac{u}{2}\sqrt{a^2-u^2} + \dfrac{a^2}{2}\sin^{-1}\dfrac{u}{a} + C$

31. $\int u^2\sqrt{a^2-u^2}\, du = \dfrac{u}{8}(2u^2-a^2)\sqrt{a^2-u^2} + \dfrac{a^4}{8}\sin^{-1}\dfrac{u}{a} + C$

32. $\int \dfrac{\sqrt{a^2-u^2}}{u}\, du = \sqrt{a^2-u^2} - a\ln\left|\dfrac{a+\sqrt{a^2-u^2}}{u}\right| + C$

33. $\int \dfrac{\sqrt{a^2-u^2}}{u^2}\, du = -\dfrac{1}{u^2}\sqrt{a^2-u^2} - \sin^{-1}\dfrac{u}{a} + C$

TABLE OF INTEGRALS

FORMS INVOLVING $\sqrt{a^2-u^2}, a>0$

34. $\displaystyle\int \frac{u^2\,du}{\sqrt{a^2-u^2}} = -\frac{u}{2}\sqrt{a^2-u^2} + \frac{a^2}{2}\sin^{-1}\frac{u}{a} + C$

35. $\displaystyle\int \frac{du}{u\sqrt{a^2-u^2}} = -\frac{1}{a}\ln\left|\frac{a+\sqrt{a^2-u^2}}{u}\right| + C$

36. $\displaystyle\int \frac{du}{u^2\sqrt{a^2-u^2}} = -\frac{1}{a^2 u}\sqrt{a^2-u^2} + c$

37. $\displaystyle\int (a^2-u^2)^{3/2}\,du = -\frac{u}{8}(2u^2-5a^2)\sqrt{a^2-u^2} + \frac{3a^4}{8}\sin^{-1}\frac{u}{a} + C$

38. $\displaystyle\int \frac{du}{(a^2-u^2)^{3/2}} = \frac{u}{a^2\sqrt{a^2-u^2}} + C$

FORMS INVOLVING $\sqrt{u^2-a^2}, a>0$

39. $\displaystyle\int \sqrt{u^2-a^2}\,du = \frac{u}{2}\sqrt{u^2-a^2} - \frac{a^2}{2}\ln\left|u+\sqrt{u^2-a^2}\right| + C$

40. $\displaystyle\int u^2\sqrt{u^2-a^2}\,du = \frac{u}{8}(2u^2-a^2)\sqrt{u^2-a^2} - \frac{a^4}{8}\ln\left|u+\sqrt{u^2-a^2}\right| + C$

41. $\displaystyle\int \frac{\sqrt{u^2-a^2}}{u}\,du = \sqrt{u^2-a^2} - a\cos^{-1}\frac{a}{|u|} + C$

42. $\displaystyle\int \frac{\sqrt{u^2-a^2}}{u^2}\,du = -\frac{\sqrt{u^2-a^2}}{u} + \ln\left|u+\sqrt{u^2-a^2}\right| + C$

43. $\displaystyle\int \frac{du}{\sqrt{u^2-a^2}} = \ln\left|u+\sqrt{u^2-a^2}\right| + C$

44. $\displaystyle\int \frac{u^2\,du}{\sqrt{u^2-a^2}} = \frac{u}{2}\sqrt{u^2-a^2} + \frac{a^2}{2}\ln\left|u+\sqrt{u^2-a^2}\right| + C$

45. $\displaystyle\int \frac{du}{u^2\sqrt{u^2-a^2}} = \frac{\sqrt{u^2-a^2}}{a^2 u} + C$

46. $\displaystyle\int \frac{du}{(u^2-a^2)^{3/2}} = -\frac{u}{a^2\sqrt{u^2-a^2}} + C$

TABLE OF INTEGRALS

FORMS INVOLVING $a + bu$

47. $\displaystyle\int \frac{u\,du}{a+bu} = \frac{1}{b^2}\left(a+bu - a\ln|a+bu|\right) + C$

48. $\displaystyle\int \frac{u^2\,du}{a+bu} = \frac{1}{2b^3}\left[(a+bu)^2 - 4a(a+bu) + 2a^2\ln|a+bu|\right] + C$

49. $\displaystyle\int \frac{du}{u(a+bu)} = \frac{1}{a}\ln\left|\frac{u}{a+bu}\right| + c$

50. $\displaystyle\int \frac{du}{u^2(a+bu)} = -\frac{1}{au} + \frac{b}{a^2}\ln\left|\frac{a+bu}{u}\right| + C$

51. $\displaystyle\int \frac{u\,du}{(a+bu)^2} = \frac{a}{b^2(a+bu)} + \frac{1}{b^2}\ln|a+bu| + C$

52. $\displaystyle\int \frac{du}{u(a+bu)^2} = \frac{1}{a(a+bu)} - \frac{1}{a^2}\ln\left|\frac{a+bu}{u}\right| + C$

53. $\displaystyle\int \frac{u^2\,du}{(a+bu)^2} = \frac{1}{b^3}\left(a+bu - \frac{a^2}{a+bu} - 2a\ln|a+bu|\right) + C$

54. $\displaystyle\int u\sqrt{a+bu}\,du = \frac{2}{15b^2}(3bu - 2a)(a+bu)^{3/2} + C$

55. $\displaystyle\int \frac{u\,du}{\sqrt{a+bu}} = \frac{2}{3b^2}(bu - 2a)\sqrt{a+bu} + C$

56. $\displaystyle\int \frac{u^2\,du}{\sqrt{a+bu}} = \frac{2}{15b^3}\left(8a^2 + 3b^2u^2 - 4abu\right)\sqrt{a+bu} + C$

57. $\displaystyle\int \frac{du}{u\sqrt{a+bu}} = \frac{1}{\sqrt{a}}\ln\left|\frac{\sqrt{a+bu} - \sqrt{a}}{\sqrt{a+bu} + \sqrt{a}}\right| + C,\ \text{if}\ a > 0$

$\qquad\qquad\qquad = \frac{2}{\sqrt{-a}}\tan^{-1}\sqrt{\frac{a+bu}{-a}} + C,\ \text{if}\ a < 0$

58. $\displaystyle\int \frac{\sqrt{a+bu}}{u}\,du = 2\sqrt{a+bu} + a\int \frac{du}{u\sqrt{a+bu}}$

59. $\displaystyle\int \frac{\sqrt{a+bu}}{u^2}\,du = -\frac{\sqrt{a+bu}}{u} + \frac{b}{2}\int \frac{du}{u\sqrt{a+bu}}$

60. $\displaystyle\int u^n\sqrt{a+bu}\,du = \frac{2}{b(2n+3)}\left[u^n(a+bu)^{3/2} - na\int u^{n-1}\sqrt{a+bu}\,du\right]$

Section 6.4: Integration with Tables

TABLE OF INTEGRALS

FORMS INVOLVING $a + bu$

61. $\int \dfrac{u^n\,du}{\sqrt{a+bu}} = \dfrac{2u^n\sqrt{a+bu}}{b(2n+1)} - \dfrac{2na}{b(2n+1)} \int \dfrac{u^{n-1}\,du}{\sqrt{a+bu}}$

62. $\int \dfrac{du}{u^n\sqrt{a+bu}} = -\dfrac{\sqrt{a+bu}}{a(n-1)u^{n-1}} - \dfrac{b(2n-3)}{2a(n-1)} \int \dfrac{du}{u^{n-1}\sqrt{a+bu}}$

TRIGONOMETRIC FORMS

63. $\int \sin^2 u\,du = \tfrac{1}{2}u - \tfrac{1}{4}\sin 2u + C$

64. $\int \cos^2 u\,du = \tfrac{1}{2}u + \tfrac{1}{4}\sin 2u + C$

65. $\int \tan^2 u\,du = \tan u - u + C$

66. $\int \cot^2 u\,du = -\cot u - u + C$

67. $\int \sin^3 u\,du = -\tfrac{1}{3}\left(2 + \sin^2 u\right)\cos u + C$

68. $\int \cos^3 u\,du = \tfrac{1}{3}\left(2 + \cos^2 u\right)\sin u + C$

69. $\int \tan^3 u\,du = \tfrac{1}{2}\tan^2 u + \ln|\cos u| + C$

70. $\int \cot^3 u\,du = -\tfrac{1}{2}\cot^2 u - \ln|\sin u| + C$

71. $\int \sec^3 u\,du = \tfrac{1}{2}\sec u \tan u + \tfrac{1}{2}\ln|\sec u + \tan u| + C$

72. $\int \csc^3 u\,du = -\tfrac{1}{2}\csc u \cot u + \tfrac{1}{2}\ln|\csc u - \cot u| + C$

73. $\int \sin^n u\,du = -\dfrac{1}{n}\sin^{n-1} u \cos u + \dfrac{n-1}{n} \int \sin^{n-2} u\,du$

74. $\int \cos^n u\,du = \dfrac{1}{n}\cos^{n-1} u \sin u + \dfrac{n-1}{n} \int \cos^{n-2} u\,du$

75. $\int \tan^n u\,du = \dfrac{1}{n-1}\tan^{n-1} u - \int \tan^{n-2} u\,du$

76. $\int \cot^n u\,du = \dfrac{-1}{n-1}\cot^{n-1} u - \int \cot^{n-2} u\,du$

77. $\int \sec^n u\,du = \dfrac{1}{n-1}\tan u \sec^{n-2} u + \dfrac{n-2}{n-1} \int \sec^{n-2} u\,du$

TABLE OF INTEGRALS

TRIGONOMETRIC FORMS

78. $\int \csc^n u \, du = \dfrac{-1}{n-1} \cot u \, \csc^{n-2} u + \dfrac{n-2}{n-1} \int \csc^{n-2} u \, du$

79. $\int \sin au \sin bu \, du = \dfrac{\sin(a-b)u}{2(a-b)} - \dfrac{\sin(a+b)u}{2(a+b)} + C$

80. $\int \cos au \cos bu \, du = \dfrac{\sin(a-b)u}{2(a-b)} + \dfrac{\sin(a+b)u}{2(a+b)} + C$

81. $\int \sin au \cos bu \, du = -\dfrac{\cos(a-b)u}{2(a-b)} - \dfrac{\cos(a+b)u}{2(a+b)} + C$

82. $\int u \sin u \, du = \sin u - u \cos u + C$

83. $\int u \cos u \, du = \cos u + u \sin u + C$

84. $\int u^n \sin u \, du = -u^n \cos u + n \int u^{n-1} \cos u \, du$

85. $\int u^n \cos u \, du = u^n \sin u - n \int u^{n-1} \sin u \, du$

86. $\int \sin^n u \cos^m u \, du = -\dfrac{\sin^{n-1} u \cos^{m+1} u}{n+m} + \dfrac{n-1}{n+m} \int \sin^{n-2} u \cos^m u \, du$

    $ = \dfrac{\sin^{n+1} u \cos^{m-1} u}{n+m} + \dfrac{m-1}{n+m} \int \sin^n u \cos^{m-2} u \, du$

INVERSE TRIGONOMETRIC FORMS

87. $\int \sin^{-1} u \, du = u \sin^{-1} u + \sqrt{1-u^2} + C$

88. $\int \cos^{-1} u \, du = u \cos^{-1} u - \sqrt{1-u^2} + C$

89. $\int \tan^{-1} u \, du = u \tan^{-1} u - \tfrac{1}{2} \ln(1+u^2) + C$

90. $\int u \sin^{-1} u \, du = \dfrac{2u^2 - 1}{4} \sin^{-1} u + \dfrac{u\sqrt{1-u^2}}{4} + C$

91. $\int u \cos^{-1} u \, du = \dfrac{2u^2 - 1}{4} \cos^{-1} u - \dfrac{u\sqrt{1-u^2}}{4} + C$

92. $\int u \tan^{-1} u \, du = \dfrac{u^2 + 1}{2} \tan^{-1} u - \dfrac{u}{2} + C$

TABLE OF INTEGRALS

INVERSE TRIGONOMETRIC FORMS

93. $\int u^n \sin^{-1} u \, du = \dfrac{1}{n+1}\left[u^{n+1} \sin^{-1} u - \int \dfrac{u^{n+1} du}{\sqrt{1-u^2}} \right], n \neq -1$

94. $\int u^n \cos^{-1} u \, du = \dfrac{1}{n+1}\left[u^{n+1} \cos^{-1} u + \int \dfrac{u^{n+1} du}{\sqrt{1-u^2}} \right], n \neq -1$

95. $\int u^n \tan^{-1} u \, du = \dfrac{1}{n+1}\left[u^{n+1} \tan^{-1} u - \int \dfrac{u^{n+1} du}{1+u^2} \right], n \neq -1$

EXPONENTIAL AND LOGARITHMIC FORMS

96. $\int u e^{au} \, du = \dfrac{1}{a^2}(au - 1)e^{au} + C$

97. $\int u^n e^{au} \, du = \dfrac{1}{a} u^n e^{au} - \dfrac{n}{a} \int u^{n-1} e^{au} \, du$

98. $\int e^{au} \sin bu \, du = \dfrac{e^{au}}{a^2 + b^2}(a \sin bu - b \cos bu) + C$

99. $\int e^{au} \cos bu \, du = \dfrac{e^{au}}{a^2 + b^2}(a \cos bu - b \sin bu) + C$

100. $\int \ln u \, du = u \ln u - u + C$

101. $\int u^n \ln u \, du = \dfrac{u^{n+1}}{(n+1)^2}\left[(n+1) \ln u - 1 \right] + C$

102. $\int \dfrac{1}{u \ln u} \, du = \ln |\ln u| + C$

HYPERBOLIC FORMS

103. $\int \sinh u \, du = \cosh u + C$

104. $\int \cosh u \, du = \sinh u + C$

105. $\int \tanh u \, du = \ln \cosh u + C$

106. $\int \coth u \, du = \ln |\sinh u| + C$

107. $\int \operatorname{sech} u \, du = \tan^{-1} |\sinh u| + C$

TABLE OF INTEGRALS

HYPERBOLIC FORMS

108. $\int \operatorname{csch} u \, du = \ln|\tanh \tfrac{1}{2} u| + C$

109. $\int \operatorname{sech}^2 u \, du = \tanh u + C$

110. $\int \operatorname{csch}^2 u \, du = -\coth u + C$

111. $\int \operatorname{sech} u \tanh u \, du = -\operatorname{sech} u + C$

112. $\int \operatorname{csch} u \coth u \, du = -\operatorname{csch} u + C$

FORMS INVOLVING $\sqrt{2au - u^2}$, $a > 0$

113. $\int \sqrt{2au - u^2} \, du = \dfrac{u-a}{2} \sqrt{2au - u^2} + \dfrac{a^2}{2} \cos^{-1}\left(\dfrac{a-u}{a}\right) + C$

114. $\int u\sqrt{2au - u^2} \, du = \dfrac{2u^2 - au - 3a^2}{6} \sqrt{2au - u^2} + \dfrac{a^3}{2} \cos^{-1}\left(\dfrac{a-u}{a}\right) + C$

115. $\int \dfrac{\sqrt{2au - u^2}}{u} \, du = \sqrt{2au - u^2} + a \cos^{-1}\left(\dfrac{a-u}{a}\right) + C$

116. $\int \dfrac{\sqrt{2au - u^2}}{u^2} \, du = -\dfrac{2\sqrt{2au - u^2}}{u} - \cos^{-1}\left(\dfrac{a-u}{a}\right) + C$

117. $\int \dfrac{du}{\sqrt{2au - u^2}} = \cos^{-1}\left(\dfrac{a-u}{a}\right) + C$

118. $\int \dfrac{u \, du}{\sqrt{2au - u^2}} = -\sqrt{2au - u^2} + a \cos^{-1}\left(\dfrac{a-u}{a}\right) + C$

119. $\int \dfrac{u^2 \, du}{\sqrt{2au - u^2}} = -\dfrac{(u+3a)}{2} \sqrt{2au - u^2} + \dfrac{3a^2}{2} \cos^{-1}\left(\dfrac{a-u}{a}\right) + C$

120. $\int \dfrac{du}{u\sqrt{2au - u^2}} = -\dfrac{\sqrt{2au - u^2}}{au} + C$

Section 6.4: Integration with Tables

Examples

1. $\int (25-4x^2)^{\frac{3}{2}} \, dx$

2. $\int \dfrac{1}{\sqrt{3+9x^2}} \, dx$

3. $\int \dfrac{1}{x^2+2x+10} \, dx$

Section 6.4: Integration with Tables

WeBWorK

2. Use the Table of Integrals to evaluate $\displaystyle\int \frac{11\,dx}{x^2\sqrt{16x^2+49}}$

Perform the substitution $u =$ _____   Use formula number _____

4. Use the Table of Integrals to evaluate $\displaystyle\int \frac{\tan^3\left(\frac{4}{z}\right)}{z^2}\,dz$

Perform the substitution $u =$ _____   Use formula number _____

Section 6.4: Integration with Tables

7. Use the Table of Integrals to evaluate $\int \dfrac{e^x}{81-e^{2x}}\,dx$

Perform the substitution $u =$ _____   Use formula number _____

Approximation

Section 6.5

Before Class Video Examples

1. Use $n = 4$ and the Trapezoidal Rule to approximate the integral $\int_1^3 \frac{5}{x^2} dx$.

2. First find the exact value of the integral $\int_1^3 \frac{5}{x^2} dx$. Then compute the error E_T that resulted from approximating this integral using the Trapezoidal Rule in the last problem.

Section 6.5: Approximation

3. Use $n = 4$ and Simpson's Rule to approximate the integral $\int_0^1 e^{x^2} dx$.

4. First, use a calculator to find the exact value of the integral $\int_0^1 e^{x^2} dx$. Then compute the error E_S that resulted from approximating this integral using Simpson's Rule in the last problem.

Algebra Review

1. **Absolute Value Functions**

   The graph of the absolute value of a function $|f(x)|$ is the reflection of the original function $f(x)$ over the x-axis and into the first and second quadrants.

   $$f(x) = x^2 - 2 \quad f(x) = |x^2 - 2|$$

   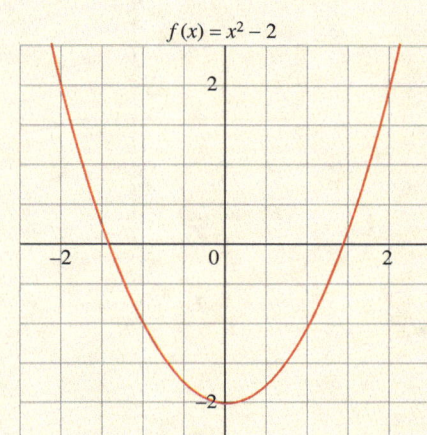
   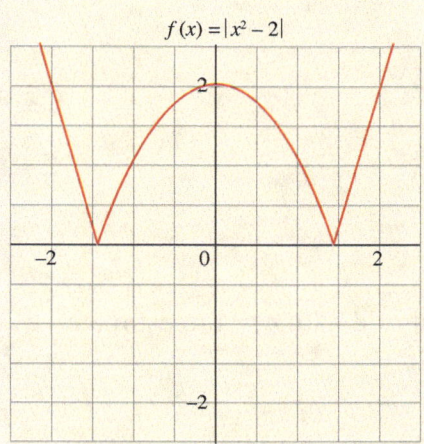

   Examples

   i. Graph $f(x) = \left|\dfrac{1}{x^3}\right|$

   ii. Give the absolute maximum value of $f(x) = \left|\dfrac{1}{x^3}\right|$ on the interval $[-3, -1]$

92 Section 6.5: Approximation

Consider the function $f(x) = e^{x^2}$.

This function cannot be integrated. However, the integral can be estimated by geometric means.

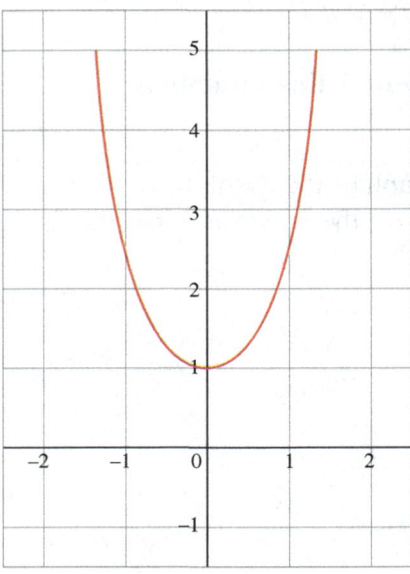

Use the midpoint rule with $n = 4$ to estimate $\int_0^1 e^{x^2}\, dx$

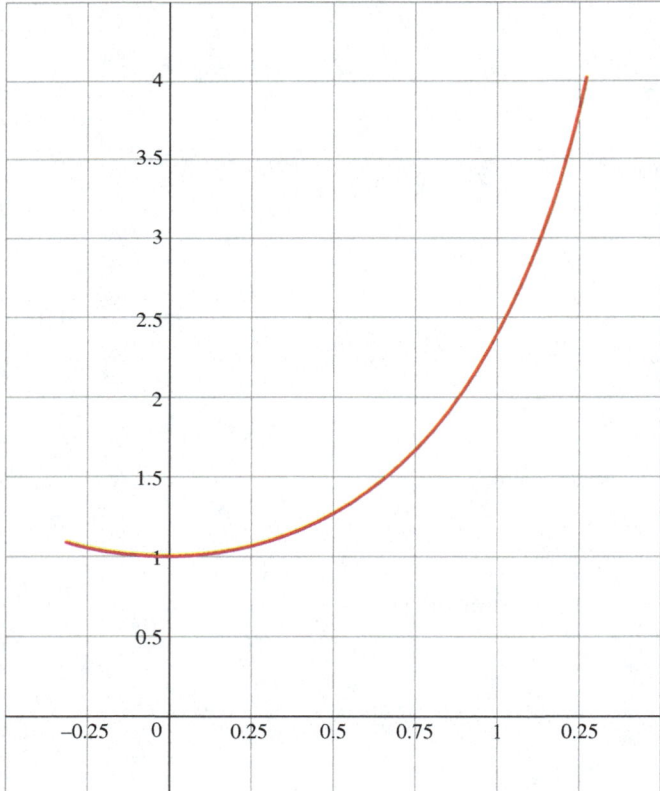

Section 6.5: Approximation 93

A The Trapezoidal Rule

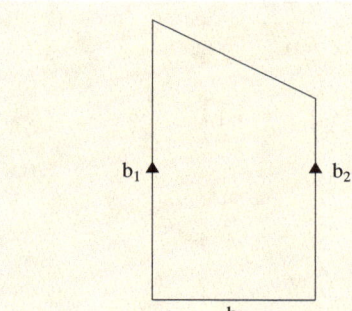

Area of a Trapezoid $= \dfrac{(b_1 + b_2) \cdot h}{2}$

Examples

1. Estimate $\int_0^1 e^{x^2} dx$ by dividing the interval $[0,1]$ into four trapezoids of equal width.

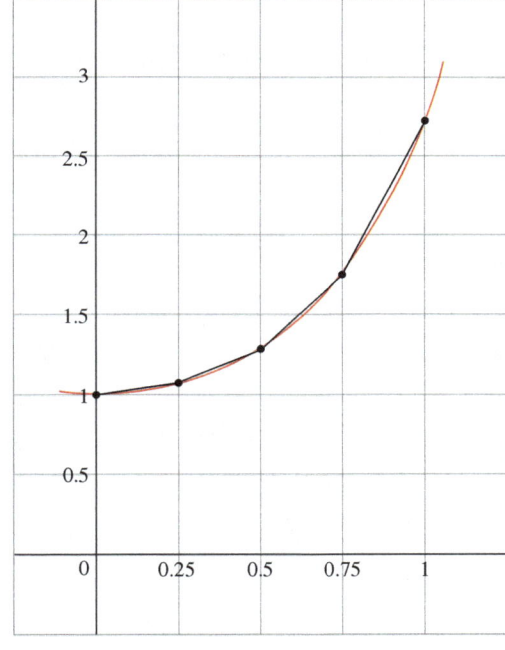 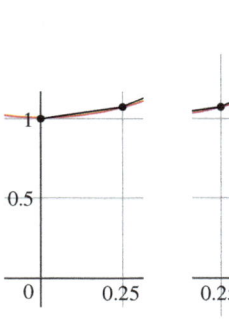 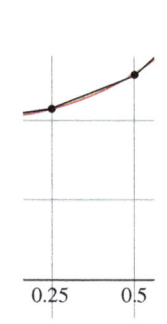 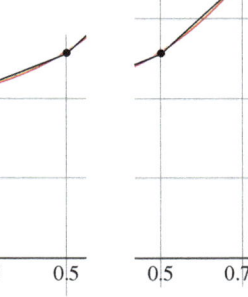 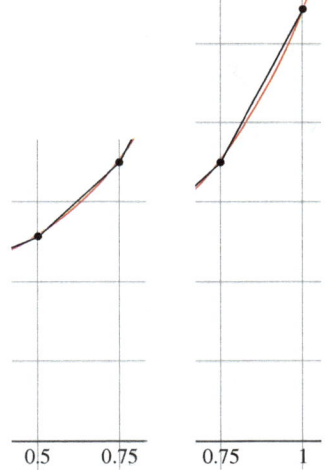

Section 6.5: Approximation

The Trapezoidal Rule can be given as:

The estimate of $\int_a^b f(x)\,dx$ Where $a = x_0$, $b = x_n$, $\Delta x = \dfrac{b-a}{n}$.

$$T_n = \dfrac{\Delta x}{2}\left[f(a) + 2f(x_1) + 2f(x_2) + \ldots + 2f(x_{n-2}) + 2f(x_{n-1}) + f(b)\right]$$

B. Simpson's Rule

Simpson's Rule is a combination of the Midpoint rule and the Trapezoidal rule. It estimates

$\int_a^b f(x)\,dx$ where $a = x_0$, $b = x_n$, $\Delta x = \dfrac{b-a}{n}$ and n is EVEN.

$$S_n = \dfrac{\Delta x}{3}\left[f(a) + 4f(x_1) + 2f(x_2) + 4f(x_3) + 2f(x_4) + \ldots + 2f(x_{n-2}) + 4f(x_{n-1}) + f(b)\right]$$

Example

2. Use Simpson's rule to estimate $\int_0^1 e^{x^2}\,dx$ with $n = 6$. (i.e., find S_6)

To summarize:

$$\text{The estimate of } \int_a^b f(x)\,dx, \text{ where } a = x_0, b = x_n \text{ and } \Delta x = \frac{b-a}{n} \text{ is given by}$$

$$L_n = \Delta x\left[f(a) + f(x_1) + f(x_2) + \ldots + f(x_{n-2}) + f(x_{n-1})\right]$$

$$R_n = \Delta x\left[f(x_1) + f(x_2) + \ldots + f(x_{n-2}) + f(x_{n-1}) + f(b)\right]$$

$$M_n = \Delta x\left[f(\overline{x}_1) + f(\overline{x}_2) + \ldots + f(\overline{x}_{n-2}) + f(\overline{x}_{n-1}) + f(\overline{x}_n)\right]$$

$$T_n = \frac{\Delta x}{2}\left[f(a) + 2f(x_1) + 2f(x_2) + \ldots + 2f(x_{n-2}) + 2f(x_{n-1}) + f(b)\right]$$

$$S_n = \frac{\Delta x}{3}\left[f(a) + 4f(x_1) + 2f(x_2) + 4f(x_3) + 2f(x_4) + \ldots + 2f(x_{n-2}) + 4f(x_{n-1}) + f(b)\right]$$

C. Error Bounds

Error bounds allow us to calculate the maximum amount of error between an approximation and the true integral for a set partition (n).

$$\text{Midpoint Rule } |E_M| \leq \frac{k \cdot (b-a)^3}{24n^2} \text{ where } |f''(x)| \leq k \text{ for } a \leq x \leq b$$

$$\text{Trapezoidal Rule } |E_T| \leq \frac{k \cdot (b-a)^3}{12n^2} \text{ where } |f''(x)| \leq k \text{ for } a \leq x \leq b$$

$$\text{Simpson's Rule } |E_S| \leq \frac{k \cdot (b-a)^5}{180n^4} \text{ where } |f^{(4)}(x)| \leq k \text{ for } a \leq x \leq b$$

Example
3. For the function $f(x) = e^{x^2}$, find

   a. $\int_0^1 e^{x^2}\,dx =$

   b. $T_4 =$

Section 6.5: Approximation

c. Find the maximum error using the error bounds and compare to the actual error.

d. What is the smallest value of n that will yield an error within 0.001?

4. For the function $f(x) = \ln(x)$, find

   a. $\displaystyle\int_1^4 \ln x \, dx$

   b. $S_6 =$

c. Find the maximum error using the error bounds and compare to the actual error.

d. What is the smallest value of n that will yield an error within 0.0001?

Section 6.5: Approximation

WeBWorK

7. A student is speeding down Highway 16 in her fancy red Porsche when her radar system warns her of an obstacle 400 ft ahead. She immediately applies the brakes, starts to slow down, and spots a skunk in the road directly ahead of her. The "black box" in the Porsche records the car's speed every 2 seconds, producing the following table. The speed decreases throughout the 10 seconds it takes to stop, although not necessarily at a uniform rate.

Time since brakes applied (s)	0	2	4	6	8	10
Speed (ft/s.)	95	85	45	25	15	0

A. What is your best estimate of the total distance the student's car traveled before coming to rest (note that the best estimate is probably not the over or under estimate that you can most easily find, use the trapezoidal approximation)?

distance = _____

B. Given the fact that the Prosche slows down during breaking, give a sharp

  i. *underestimate* of the distance traveled:_____

  ii. *overestimate* of the distance traveled: _____

C. Which one of the following statements can you justify from the information given?

- A. The "black box" data is inconclusive. The skunk may or may not have been hit.
- B. The skunk was hit by the car.
- C. The car stopped before getting to the skunk.

Improper Integrals

Section 6.6

Before Class Video Examples

1. Determine whether the improper converges or diverges integral $\int_{3}^{\infty} \frac{5}{x} dx$ and evaluate the integral if is convergent.

2. Determine whether the improper integral $\int_{3}^{\infty} \frac{5}{x^2} dx$ and evaluate the integral if is convergent.

Section 6.6: Improper Integrals

3. Determine whether the improper integral $\int_{2}^{11} \dfrac{5}{\sqrt{x-2}}\,dx$ and evaluate the integral if is convergent.

4. Determine whether the improper integral $\int_{3}^{10} \dfrac{7}{x-5}\,dx$ and evaluate the integral if is convergent.

Section 6.6: Improper Integrals 101

Algebra Review

Shapes of Graphs

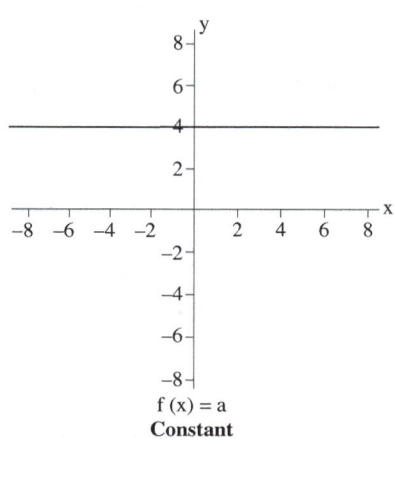
$f(x) = a$
Constant

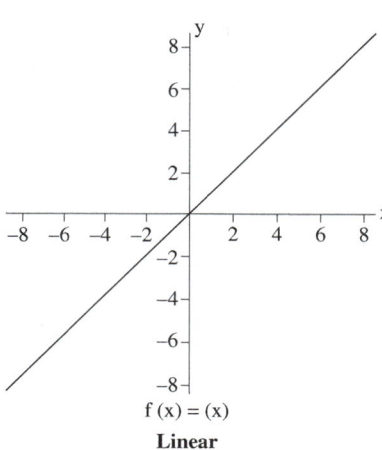
$f(x) = (x)$
Linear

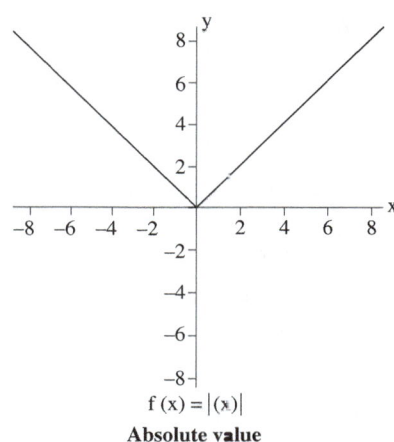
$f(x) = |(x)|$
Absolute value

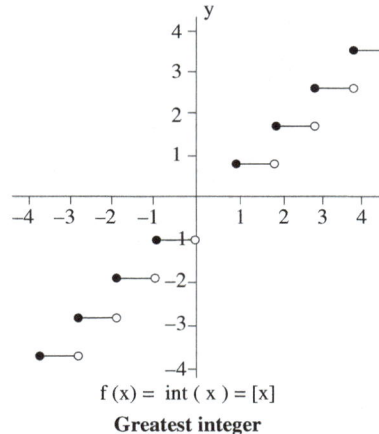
$f(x) = \text{int}(x) = [x]$
Greatest integer

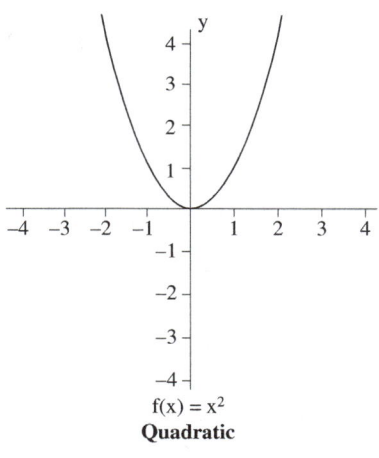
$f(x) = x^2$
Quadratic

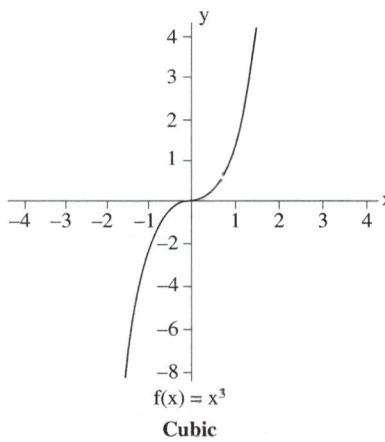
$f(x) = x^3$
Cubic

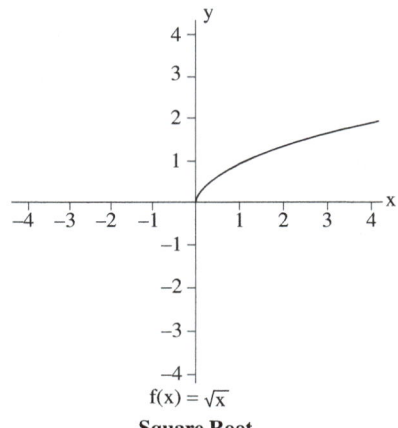
$f(x) = \sqrt{x}$
Square Root

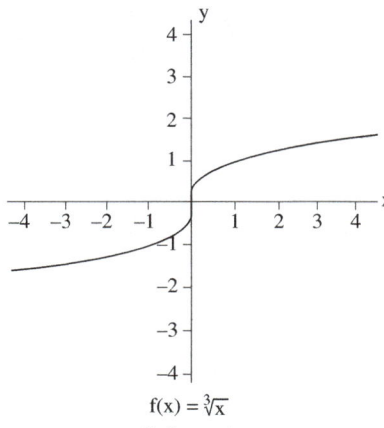
$f(x) = \sqrt[3]{x}$
Cube root

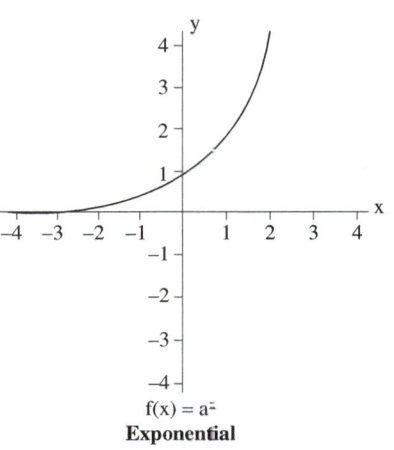
$f(x) = a^x$
Exponential

Section 6.6: Improper Integrals

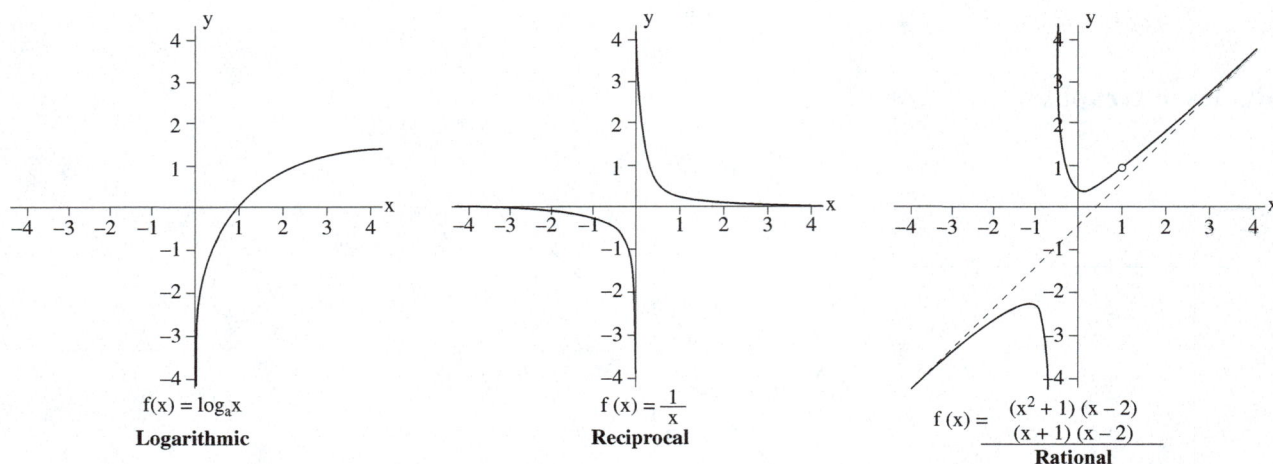

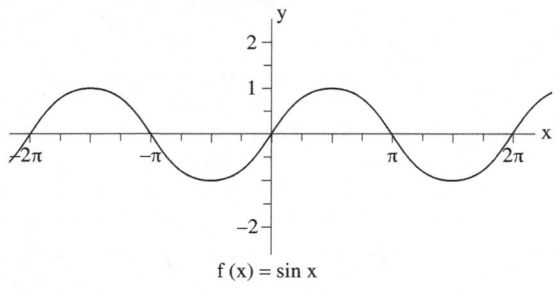

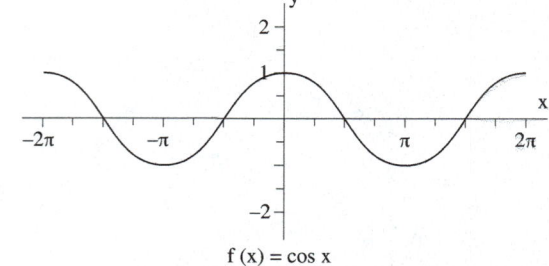

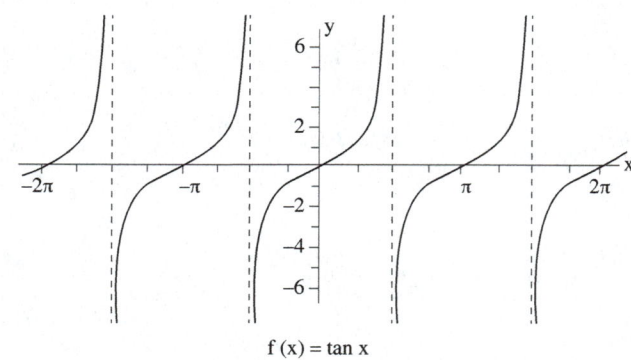

Section 6.6: Improper Integrals

A. Review of Limit Rules

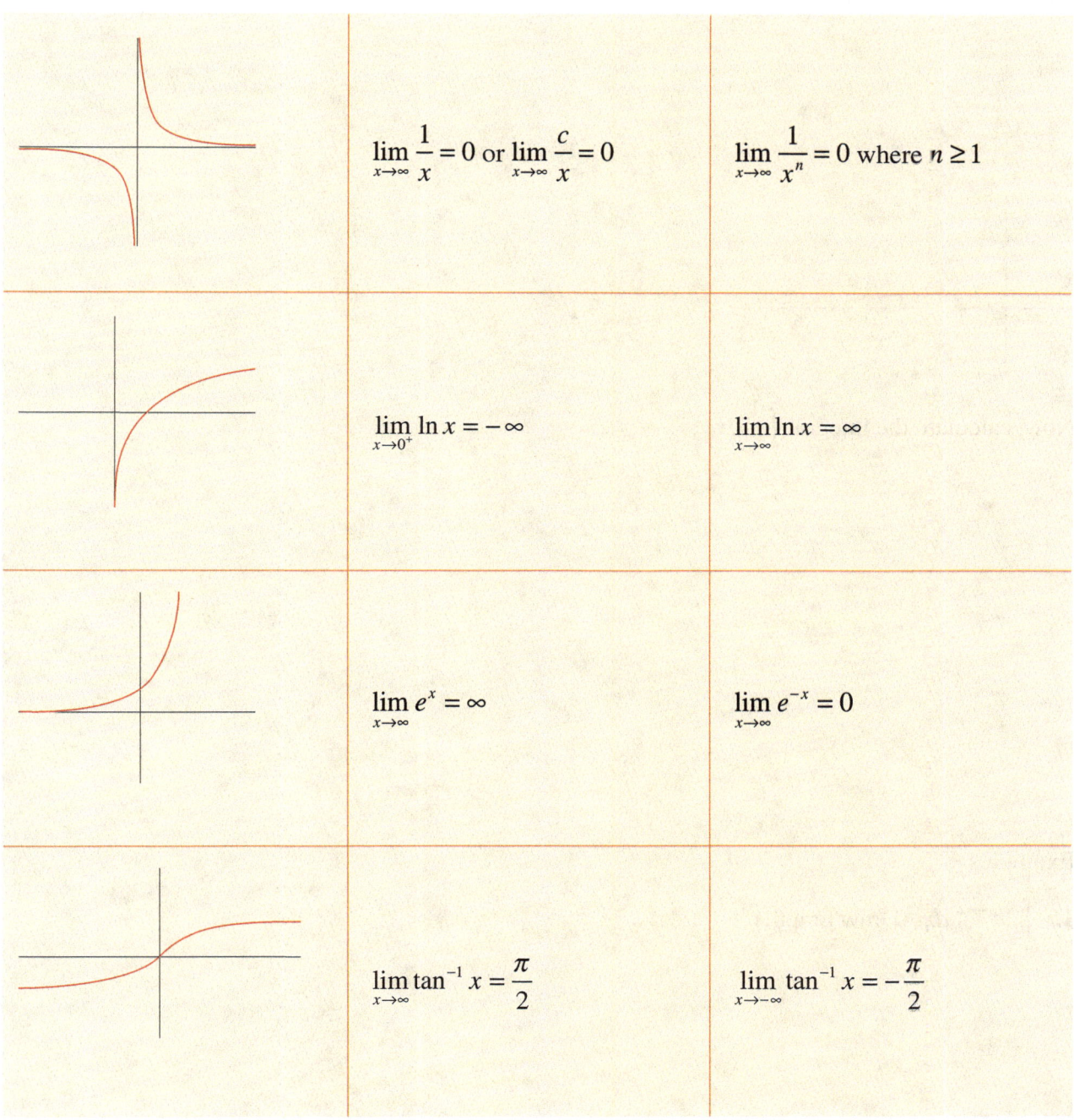

	$\lim_{x \to \infty} \dfrac{1}{x} = 0$ or $\lim_{x \to \infty} \dfrac{c}{x} = 0$	$\lim_{x \to \infty} \dfrac{1}{x^n} = 0$ where $n \geq 1$
	$\lim_{x \to 0^+} \ln x = -\infty$	$\lim_{x \to \infty} \ln x = \infty$
	$\lim_{x \to \infty} e^x = \infty$	$\lim_{x \to \infty} e^{-x} = 0$
	$\lim_{x \to \infty} \tan^{-1} x = \dfrac{\pi}{2}$	$\lim_{x \to -\infty} \tan^{-1} x = -\dfrac{\pi}{2}$

Section 6.6: Improper Integrals

Consider the function $f(x) = \dfrac{1}{x}$. Shade the area calculated by $\displaystyle\int_0^2 \dfrac{1}{x}\,dx$

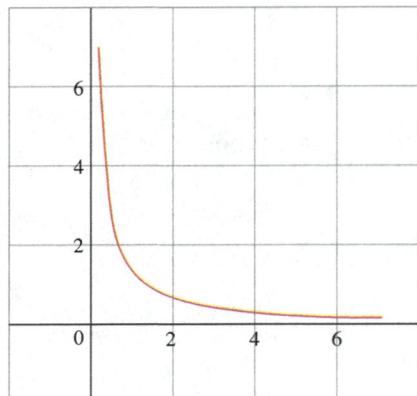

Now, calculate the integral $\displaystyle\int_0^2 \dfrac{1}{x}\,dx$

Examples

1. $\displaystyle\int_0^\infty 2e^{-4x}\,dx$. (Draw graph!)

B. Notation

To show the procedure of the limit, we will substitute the "improper" bound with a variable of our choice, and find the limit as the variable approached the limit.

Instead of $\int_0^2 \frac{1}{x} dx$ we will show $\lim_{A \to 0^+} \int_A^2 \frac{1}{x} dx$

When an integral = Constant, we say that it CONVERGES to that number.

When an integral = Infinity, we say that it DIVERGES.

Examples

2. $\int_3^\infty \frac{1}{x^2} dx$

3. $\int_3^\infty \frac{1}{x} dx$

Section 6.6: Improper Integrals

4. $\displaystyle\int_{2}^{6} \frac{1}{x-2}\, dx$

5. $\displaystyle\int_{1}^{\infty} \frac{1}{1+x^2}\, dx$

6. $\displaystyle\int_{-\infty}^{\infty} \frac{1}{r^2+4}\,dr$

Section 6.6: Improper Integrals

7. $\displaystyle\int_{-2}^{3} \frac{1}{x^4}\,dx$

C. Comparison Theorem (aka Squeeze or Sandwich Theorem)

For continuous functions $f(x)$ and $g(x)$, where $f(x) \geq g(x) \geq 0$ and $x \geq c$, we can conclude that

i. $\displaystyle\int_{c}^{\infty} f(x)\,dx$ is convergent $\Rightarrow \displaystyle\int_{c}^{\infty} g(x)\,dx$ is also convergent

ii. $\displaystyle\int_{c}^{\infty} g(x)\,dx$ is divergent $\Rightarrow \displaystyle\int_{c}^{\infty} f(x)\,dx$ is also divergent

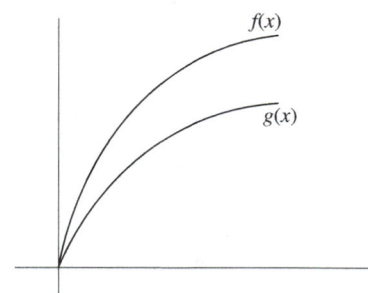 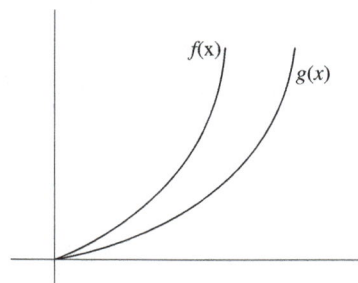

Examples

8. $\displaystyle\int_{1}^{\infty} \frac{\cos^2 x}{1+x^2}\,dx$

Section 6.6: Improper Integrals

WeBWorK

16. Enter T if the given statement is true or F if it is false.

a. $\int_1^\infty \dfrac{1}{x+e^{6x}}\, dx$ is divergent because $\dfrac{1}{x+e^{6x}} \leq \dfrac{1}{x}$ and $\int_1^\infty \dfrac{1}{x}\, dx$ is divergent.

b. $\int_1^\infty \dfrac{1}{x+e^{6x}}\, dx$ is convergent because $\dfrac{1}{x+e^{6x}} \leq \dfrac{1}{e^{6x}}$ and $\int_1^\infty \dfrac{1}{e^{6x}}\, dx$ is convergent.

c. $\int_1^\infty \dfrac{1}{x-e^{-6x}}\,dx$ is divergent because $\dfrac{1}{x-e^{-6x}} \geq \dfrac{1}{x}$ and $\int_1^\infty \dfrac{1}{x}\,dx$ is divergent.

d. $\int_1^\infty \dfrac{1}{x+e^{6x}}\,dx$ is convergent because $\dfrac{1}{x+e^{6x}} \leq \dfrac{1}{x}$ and $\int_1^\infty \dfrac{1}{x}\,dx$ is convergent.

Area Between Curves

Section 7.1

Before Class Video Examples

1. Sketch and find the area bound between the curves $y = 1$ and $y = \dfrac{1}{x^2}$ for $1 \leq x \leq 2$

2. Find the *points* of intersection for the functions $y = 6x - x^2$ and $y = x$

Section 7.1: Area Between Curves

3. Find the area bounded by the functions $y = 6x - x^2$ and $y = x$

4. Find the area bounded by the functions $y = e^x$ and $y = 0$ and the lines $x = 0$ and $x = \ln(3)$

Section 7.1: Area Between Curves   115

Algebra Review

1. **Functions in Terms of x and y**

   Functions in terms of x: $f(x) = 2x+1 \Leftrightarrow y = 2x+1$

   Functions in terms of y: $f(y) = 3y+12 \Leftrightarrow x = 3y+12 \Leftrightarrow y = \dfrac{1}{3}x - 4$

Example

i. Give $x = g(y) = \ln(y+4)$ as a function of x

ii. Give $y = f(x) = x^{\frac{3}{2}} - 1$ as a function of y

Section 7.1: Area Between Curves

A. Area Between Two Curves

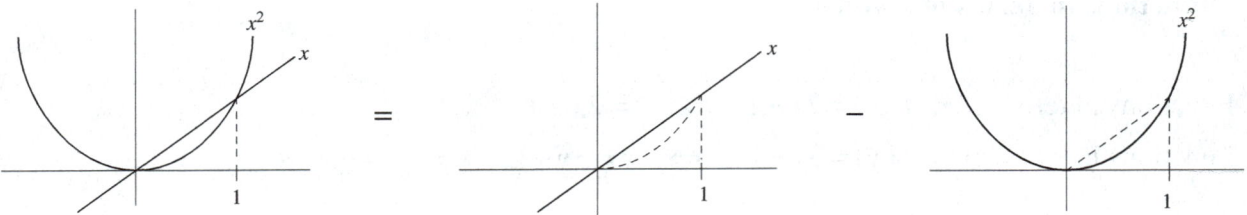

$$\text{Area} = \int_0^1 x - x^2 \, dx = \frac{x^2}{2} - \frac{x^3}{3} \Big|_0^1 = \left(\frac{1}{2} - \frac{1}{3}\right) = \frac{1}{6} \text{ unit}^2$$

$$\text{Area between two curves} = \int_A^B [\overset{\text{TOP}}{f(x)}] - [\overset{\text{BOTTOM}}{g(x)}] \, dx$$

Example

1. Sketch and find the area bound between the curves $f(x) = \sqrt{x}$ and $g(x) = \frac{x}{3}$

Section 7.1: Area Between Curves

$$\text{Area between two curves} = \int_A^B \overset{\text{RIGHT}}{[f(y)]} - \overset{\text{LEFT}}{[g(y)]} \, dy$$

Example

2. Sketch and find the area bound between the curves $y^2 = x$ and $x - 2y = 3$

Section 7.1: Area Between Curves

3. Sketch and find the area bound between the curves $f(x) = 4\sin x$ and $h(x) = 4\cos x$ on the interval $[0, \pi]$

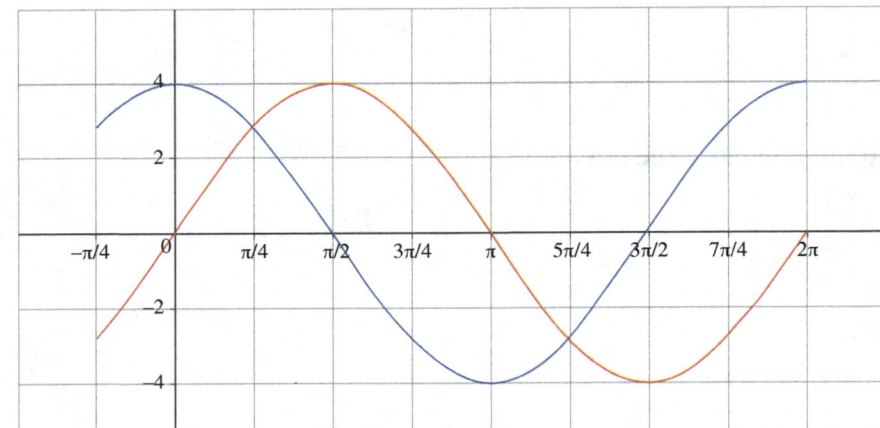

B. Area Between Multiple Curves/Functions

Example

4. Sketch and find the area that is enclosed by the graphs of $f(x)=\sqrt{x}$, $g(x)=2-x$, and $h(x)=2$

Section 7.1: Area Between Curves

1. The widths (in meters) of a kidney-shaped swimming pool, measured at 2-meter intervals, are: 0, 3.5, 5.5, 5.6, 7.2, 6.5, and 0. Use Simpson's Rule to estimate the area of the pool.

2. Find the value of $c > 0$ such that the area of the region enclosed by the parabolas $y = x^2 - c^2$ and $y = c^2 - x^2$ is 260.

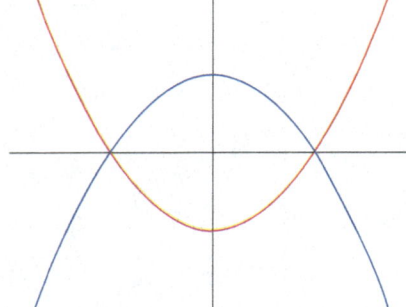

3. Find the number b such that the line $y = b$ divides the region bounded by the curves $y = x^2$ and $y = 25$ into two regions with equal area.

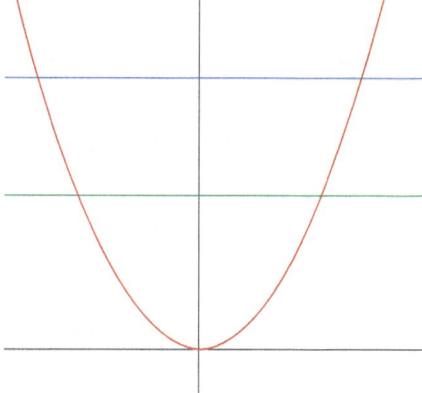

Volume

Section 7.2/7.3

Before Class Video Examples

1. Sketch and find the volume of the region in the first quadrant bounded by the curves $y = x^3$, $y = 0$, and $x = 3$ rotated about the x-axis. First set up the integral then evaluate it.

2. Sketch and find the volume of the region bounded by the curves $y = x^3$, $y = 8$, and $x = 0$ rotated about the x-axis. Just set up the integral.

3. Sketch and find the volume of the region in the first quadrant bounded by the curves $y = x^3$, $y = 0$, and $x = 3$ rotated about the line $y = -1$. Just set up the integral.

4. Sketch and find the volume of the region in the first quadrant bounded by the curves $y = x^3$ and $y = x$ rotated about the x-axis. First set up the integral then evaluate it.

5. Sketch and find the volume of the region in the first quadrant bounded by the curves $y = 4x - x^3$ and $y = 0$ rotated about the y-axis. First set up the integral then evaluate it.

   Notice that you can only use the Cylindrical Shells Method and NOT the Slicing (Disc/Washer) Method because using the Slicing (Disc/Washer) Method requires you to solve for x in terms of y to rotate around the y-axis.

6. Sketch and find the volume of the region bounded by the curves $y = \dfrac{3}{x}$, $y = 0$, with $x = 1$ and $x = 3$ rotated about the y-axis. Just set up the integral using the (Cylindrical) Shell Method.

7. Sketch and find the volume of the region bounded by the curves $y = \dfrac{3}{x}$, $y = 0$, with $x = 1$ and $x = 3$ rotated about the line $x = -1$. Just set up the integral using the (Cylindrical) Shell Method.

Section 7.2/7.3: Volume 127

Algebra Review

1. **Volume of a Cyllinder/Disc**

$$Volume = \pi R^2 H$$

2. **Volume of a Washer**

$$Volume = \pi R^2 H - \pi r^2 H = \pi\left(R^2 - r^2\right)H$$

3. **Distance Between functions**

Functions defined in terms of x

$f(x)$ and $g(x)$, where $f(x) > g(x)$ for all x the distance between f and g is given by:
Distance $= f(x) - g(x) =$ Top $-$ Bottom

Functions defined in terms of y

$f(y)$ and $g(y)$, where $f(y) > g(y)$ for all y the distance between f and g is given by:
Distance $= f(y) - g(y) =$ Right $-$ Left

Example

i. Determine the distance between $y = -x^2 + 4$ and $y = x + 1$

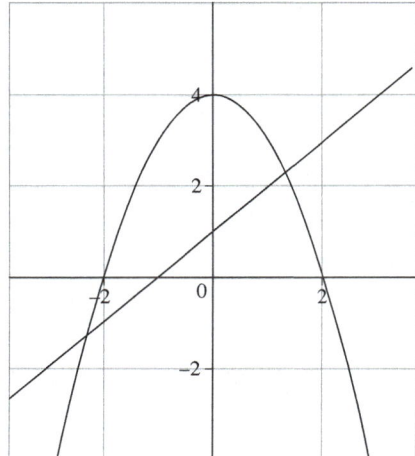

ii. Determine the distance between $x = y^2 + 2$ and $x = \dfrac{1}{2}y - 2$

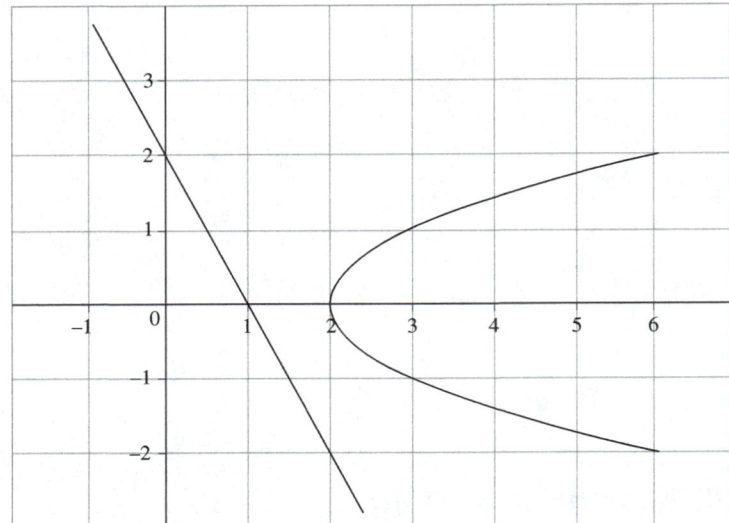

4. **Area of Equalateral Triangles**

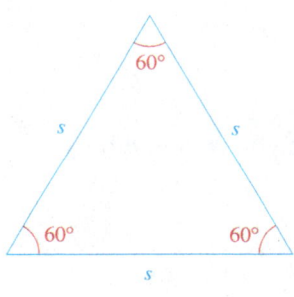

Equilateral triangle can be divided into two right triangles by the line that defines the height of the original triangle.

By Pythagorean Theorem, this height can be described as:

$$H^2 = s^2 - \left(\dfrac{s}{2}\right)^2$$

$$H = \sqrt{s^2 - \left(\dfrac{s}{2}\right)^2} = \sqrt{s^2 - \dfrac{1}{4}s^2} = \sqrt{\dfrac{3}{4}s^2} = \dfrac{\sqrt{3}}{2}s$$

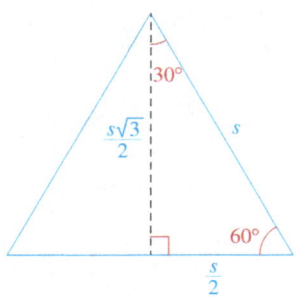

$$\text{Area} = \dfrac{1}{2}BH$$

$$\text{Area} = \dfrac{1}{2}s \cdot \dfrac{\sqrt{3}}{2}s = \dfrac{\sqrt{3}}{4}s^2$$

5. Area of Isosceles Right Triangles

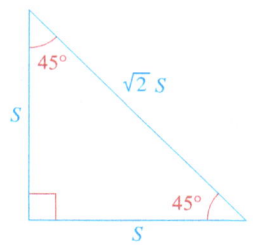

Isosceles Right triangle has base = height

$$Area = \frac{1}{2}BH = \frac{1}{2}B^2 = \frac{1}{2}H^2 = \frac{1}{2}S^2$$

6. Ellipse

Standard Form $\left(\dfrac{x}{a}\right)^2 + \left(\dfrac{y}{b}\right)^2 = 1$

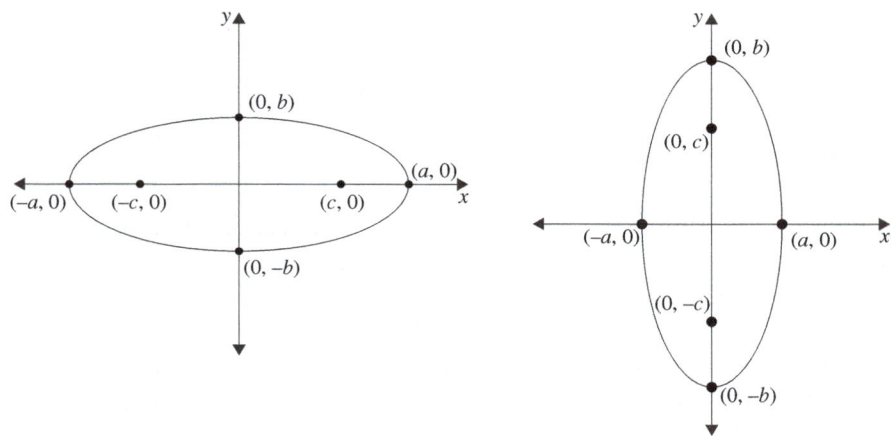

7. Geometric Shapes

Frustum of a Cone

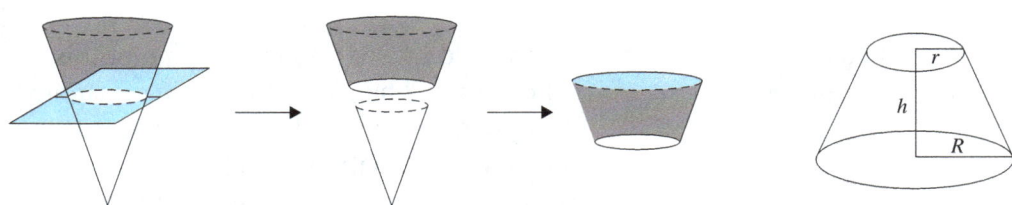

Cap of a Sphere

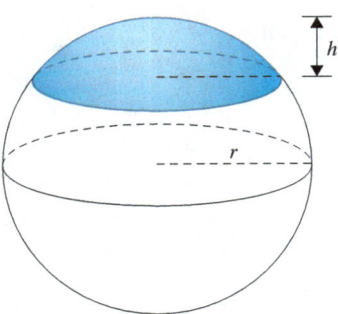

A. Disc Method

Consider the graph of a random, continuous function $f(x)$.

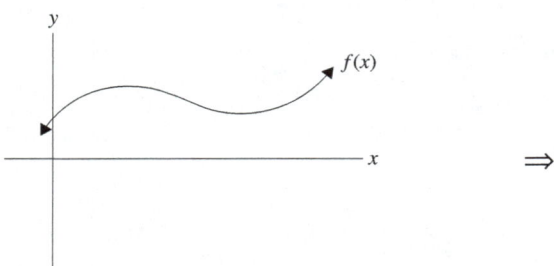

If we assume that this rotation has formed a solid object, it would look a bit like a flower pot.

After rotating the graph of $f(x)$ around the x-axis, we get

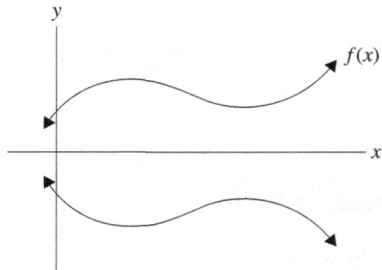

We would like to find the volume of this "flower pot," but since the shape resemble as cylinder which radius is constantly changing, we will have to break it into pieces/discs, find the volume for each disc, and add them all together for an estimate of the total volume.

Section 7.2/7.3: Volume

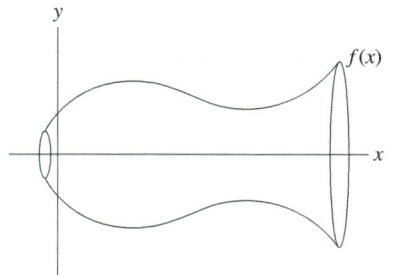 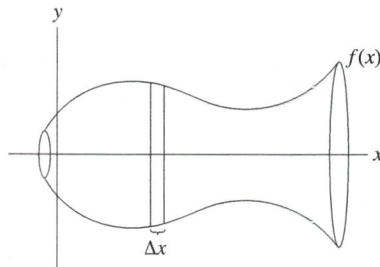

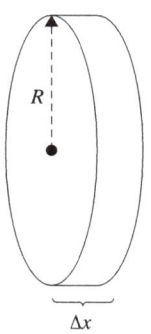

To find the volume for one of these discs, we have to treat it as a cylinder (we will make the slice thin enough so that the slight curvature to the sides is neglectable.) We will call the height of each of these discs Δx and the radius will be $f(x)$.

So,

$$V_{disc} = \pi \cdot R^2 \cdot H = \pi \cdot [f(x)]^2 \cdot \Delta x$$

Adding up all of the discs gives:
$$V = \sum_{a}^{b} \pi \cdot [f(x)]^2 \cdot \Delta x$$

To ensure the least possible error, we will have to let Δx be as small as possible, so we will find the limit as Δx approaches infinity. This yields $V = \lim\limits_{\Delta x \to 0} \sum_{a}^{b} \pi \cdot [f(x)]^2 \cdot \Delta x$. However, recall:

Definition of the Integral (from Section 5.2):
$$\int_a^b f(x)\,dx = \lim_{n \to \infty} \sum_{i=1}^{n} f(x_i) \cdot \Delta x \text{ where } \Delta x = \frac{b-a}{n} \text{ and } x_i = a + i \cdot \Delta x$$

We get that $V = \int_a^b \pi \cdot [f(x)]^2 \, dx$

Disc Method: $V = \pi \int_a^b [f(x)]^2 \, dx$

132 Section 7.2/7.3: Volume

Example

1. Sketch and find the volume that is created by rotating the area bound by the curves $y = e^x$, $y = 0$, $x = 0$, and $x = 1$ about the x-axis.

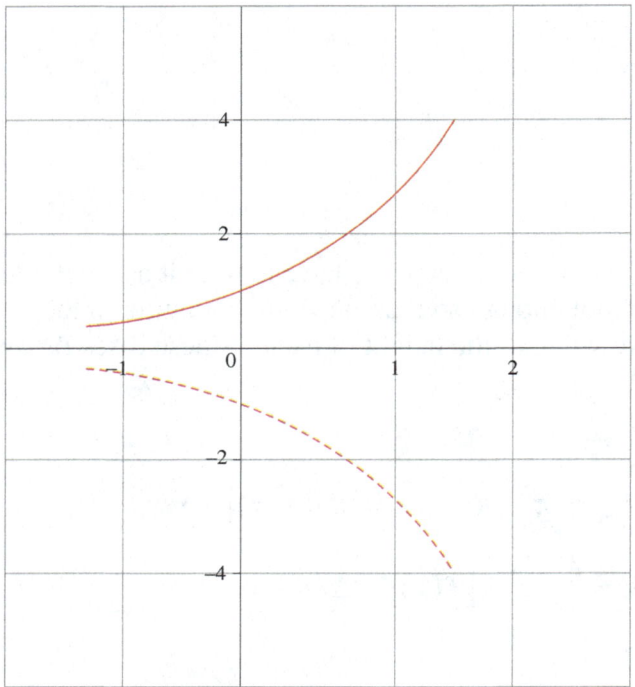

B. Washer Method

Deriving the formula for a shape that has a hollow, is a similar process. Since the hollow is caused by a second graph, we will simply subtract the "inside volume" from the "outside volume."

After rotating the graph of $f(x)$ and $g(x)$ around the x-axis, we get

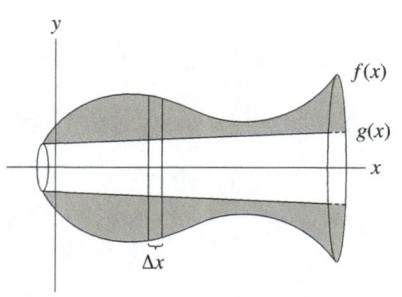

$$V_{washer} = (\pi \cdot R^2 \cdot H) - (\pi \cdot r^2 \cdot H) = \pi \cdot (R^2 - r^2) \cdot H = \pi \cdot \left([f(x)]^2 - [g(x)]^2 \right) \cdot \Delta x$$

Section 7.2/7.3: Volume

If we add up all of the discs, we will get, $V = \sum_{a}^{b} \pi \cdot \left([f(x)]^2 - [g(x)]^2 \right) \cdot \Delta x$

To ensure the least possible error, we will have to let Δx be as small as possible, so we will find the limit as Δx approaches infinity.

This yields $V = \lim_{\Delta x \to 0} \sum_{a}^{b} \pi \cdot \left([f(x)]^2 - [g(x)]^2 \right) \cdot \Delta x$ However, by

the definition of the integral, we know that $\lim \sum \Rightarrow \int$ and $\Delta x \Rightarrow dx$

So, $V = \int_{a}^{b} \pi \cdot \left([f(x)]^2 - [g(x)]^2 \right) dx$

> Washer Method: $V = \pi \int_{a}^{b} [f(x)]^2 - [g(x)]^2 \, dx$ when rotating about the x-axis

Example

2. Sketch and find the volume that is created by rotating the area bound by the curves $y = 2x - x^2$ and $y = x^2$, about the x-axis.

Section 7.2/7.3: Volume

Washer Method: $V = \pi \int_a^b [f(y)]^2 - [g(y)]^2 \, dy$ when rotating about the y-axis

Example

3. Sketch and find the volume that is created by rotating the area bound by the curves $y = x^{\frac{2}{3}}$, $y = 0$, and $x = 1$ about the y-axis.

C. Rotating around Other Axis

i. **Rotating around an axis that is parallel to the x-axis**

$$V = \pi \int_a^b \left[\text{"Big" Functional RADIUS}\right]^2 - \left[\text{"Small" Functional radius}\right]^2 dx$$

Functional RADIUS: The distance between the **axis of rotation** and the **outside/big radius**

Functional radius: The distance between the **axis of rotation** and the **inside/small radius**

Example

4. Sketch an find the volume that is created by rotating the area bound by the curves $y = \frac{1}{x}$, $y = 0$, $x = 1$, and $x = 3$ about the line $y = -1$.

ii. Rotating around an axis that is parallel to the y-axis

$$V = \pi \int_a^b \left[\text{"Big" Functional RADIUS}\right]^2 - \left[\text{"Small" Functional radius}\right]^2 dy$$

Functional RADIUS: The distance between the axis of rotation and the outside/big radius

Functional radius: The distance between the axis of rotation and the inside/little radius

Example

5. Sketch and find the volume that is created by rotating the area bound by the curves $y = x^{\frac{2}{3}}$, $y = 0$, and $x = 1$ about the line $x = 4$.

136 Section 7.2/7.3: Volume

A. Shell Method

Shell Method is used to calculate the volume when rotating about the y-axis (or an axis parallel to the y-axis).

It is especially useful when we have a functions in terms of x, but we are unable to solve for x.

Instead of creating slices that are perpendicular to the axis of rotation (discs), we will instead slice the volume into shells by viewing concentric circles around the axis of rotation.

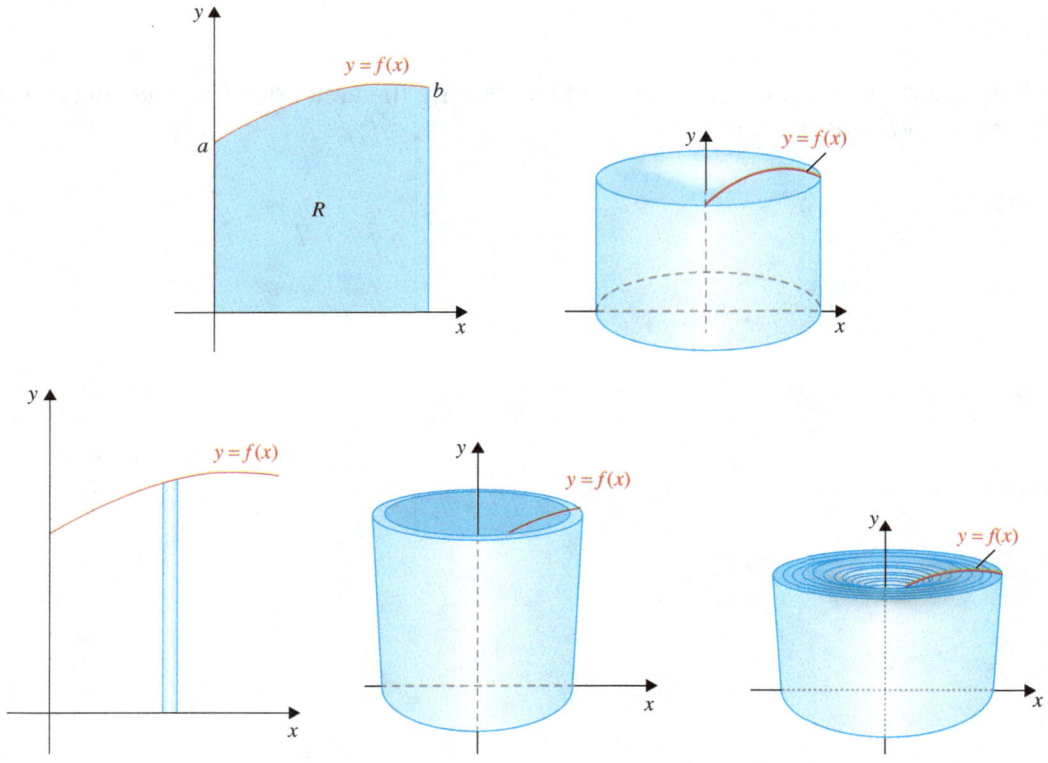

Each shell is an empty cylinder. When disassembled it is a rectangular prism with dimensions
High $= f(x)$, Width $= 2\pi x_i$ Depth $= \Delta x$

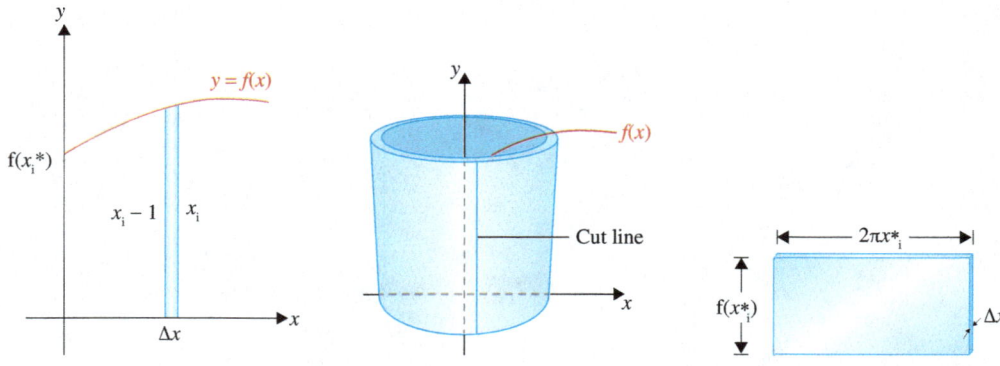

The volume for each shell is given by $H \times W \times D = f(x) \cdot 2\pi x_i \cdot \Delta x$

So,
$$V_{One\ Shell} = H \cdot W \cdot D = f(x) \cdot 2\pi x_i \cdot \Delta x$$

Adding up all of the discs gives:
$$V = \sum_a^b f(x) \cdot 2\pi x_i \cdot \Delta x$$

To ensure the least possible error, we will have to let Δx be as small as possible, so we will find the limit as Δx approaches 0.

This yields $V = \lim_{\Delta x \to 0} \sum_a^b f(x) \cdot 2\pi x_i \cdot \Delta x$. However, by the definition of the integral; we know that $\lim \sum \Rightarrow \int$ and $\Delta x \Rightarrow dx$ So, $V = \int_a^b 2\pi \cdot (x)(f(x))dx$

$$V = 2\pi \int_a^b (\text{Point Radius}) \cdot (f(x) - g(x))dx$$

- Where $f(x)$ is the function that is on TOP and $g(x)$ is the function on the BOTTOM.
- **Point Radius**: The radius of the shell given as a function of x. It can also be viewed as the distance between a random point x in the region and the axis of rotation.

Note, synonymous expressions
- Circumference of the shell = $2\pi \cdot$Point Radius
- Height of the shell = $f(x) - g(x)$

Example

6. Sketch and find the volume that is created by rotating the area bound by the curves $y = x^{\frac{2}{3}}$, $y = 0$, and $x = 1$ about the y-axis.

Section 7.2/7.3: Volume

7. Sketch and find the volume that is created by rotating the area bound by the curves $y = x^{\frac{2}{3}}$, $y = 0$, and $x = 1$ about the line $x = 4$.

8. Sketch and find the volume that is created by rotating the area bound by the curves $y = x^{\frac{2}{3}}$, $y = 0$, and $x = 1$ about the line $x = -2$.

9. Sketch and find the volume that is created by rotating the area bound by the curves $y = x^3 - 2x + 4$, $y = 0$, $x = -2$, and $x = 2$ about the line $x = -3$.

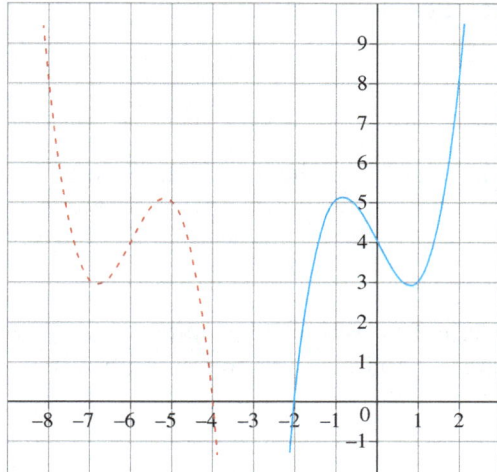

B. Nonrotational Solids

Unlike rotational solids that are formed by rotation, "nonrotational" solids are formed by accumulated cross sections.

$$V = \int_a^b Area(x)\, dx \qquad Area(x) \text{ is the function that describes the area of each slice.}$$

10. A three-dimensional solid shape has a base that is the region between the graphs of $y = x^2$ and $y = 16$. Parallel cross sections with respect to the x-axis are squares. Find the volume of the solid.

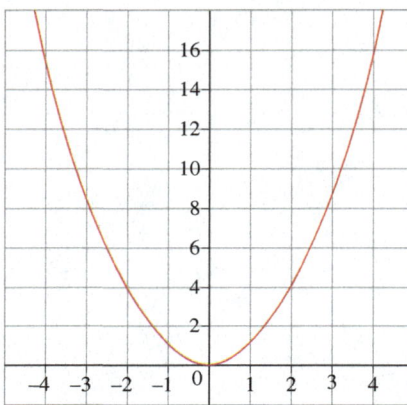

11. A three-dimensional solid shape has a base that is the region between the graphs of $y = x^2$ and $y = 16$. Parallel cross sections with respect to the y-axis are squares. Find the volume of the solid.

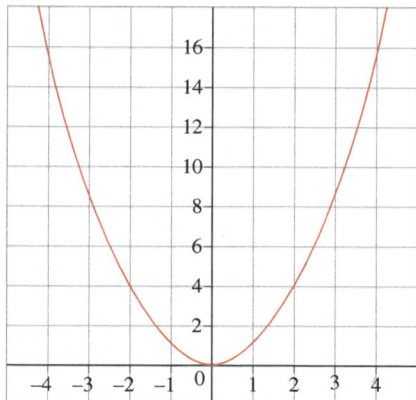

Section 7.2/7.3: Volume

12. A three-dimensional solid shape has a circular base of radius 4. Parallel cross sections (since it is a circular base, the cross sections can be made with the x or y axis) are equilateral triangles. Find the volume of the solid.

More Examples

Examples

13. Consider the region bounded be the curves $y = x$ and $y = -x^2 + 2x$

   a. Find the area of the region.

   b. Find the volume of the region rotated about the x-axis.

c. Find the volume of the region rotated about the line $y = -1$.

d. Find the volume of the region rotated about the line $y = 2$.

e. Find the volume of the region rotated about the *y*-axis.

f. Find the volume of the region rotated about the line $x = 2$.

g. Find the volume of the region rotated about the line $x = -3$.

Section 7.2/7.3: Volume

WeBWorK 7.2

8. Find the volume of the frustum of a right circular cone with height $h = 14$, lower base radius $R = 15$, and top radius $r = 12$.

9. Find the volume of a cap of a sphere with radius $r = 16$ and height $h = 5$.

6. The base of a certain solid is an elliptical region with boundary curve $9x^2 + 16y^2 = 144$. Cross sections perpendicular to the x-axis are isosceles right triangles with hypotenuse in the base.

Use the formula $V = \int_a^b A(x)\,dx$ to find the volume of the solid.

The lower limit of integration is $a =$ _____

The upper limit of integration is $b =$ _____

The base of the triangular cross section is the following function of x: _____

The height of the triangular cross section is the following function of x: _____

The area of the triangular cross section is $A(x) =$ _____

Thus the volume of the solid is $V =$ _____

Section 7.2/7.3: Volume

WeBWorK 7.3

6. Find the volume V of the solid S obtained by rotating the region bounded by the given curves about the x-axis: $y = x^3$, $x = 0$, and $y = 27$.

   A. Use the method of cylindrical shells to find the volume V of the solid S:

   $$V = \int_a^b \boxed{}\, dy.$$

   The circumference of a typical shell in terms of $y =$

   The height of this shell in terms of $y =$

   B. Use the method of slicing to find the volume V of the solid S:

   $$V = \int_a^b \boxed{}\, dx.$$

Arc Length

Section 7.4

Before Class Video Examples

1. Give the integral that gives the length of the curve $y = \sin(x)$ for $0 \leq x \leq \pi$?

2. Give the integral for the length of the curve $x = \dfrac{2}{y}$ between the points $(1, 2)$ and $\left(2, \dfrac{1}{2}\right)$.

3. Compute the arc length of the curve for $y = 4x^{\frac{3}{2}}$ for $0 \leq x \leq 1$

Section 7.4: Arc Length

Algebra Review

1. **Factoring Polynomial Quadratic Functions**

Quadratic functions with leading coefficient $=1$

Example

 i. $x^2 + 10x + 16 =$

Quadratic functions with leading coefficient $\neq 1$

Example

 ii. $8x^2 - 10x - 3 =$

2. **Factoring Quadratic "Wanna-be" Functions**

Forms Involving Higher Polynomial Functions

Example

 iii. $x^6 - 5x^3 - 14 =$

Forms Involving Exponential Functions

Example

 iv. $e^{2x} - 2e^x - 15 =$

Forms Involving Trigonometric Functions

Example

v. $\sin^2 x - \dfrac{1}{2}\sin x - \dfrac{1}{2} =$

Arc Length

Arc Length for a Function of $x = \displaystyle\int_A^B \sqrt{1+(f'(x))^2}\, dx$

Arc Length for a Function of $y = \displaystyle\int_A^B \sqrt{1+(f'(y))^2}\, dy$

Examples

1. Find the arc length of $x = y^{\frac{3}{2}}$ between $y = 0$ and $y = 1$.

Section 7.4: Arc Length

2. Find the arc length of $f(x) = \dfrac{x^2}{2} - \dfrac{\ln x}{4}$ between $x = 2$ and $x = 4$

3. Set up the integral that will calculate the arc length of $y = \cos x$ on $0 \leq x \leq 2\pi$. Use your calculator to evaluate.

Section 7.4: Arc Length

WeBWorK

2. Consider the curve defined by $y = \dfrac{x^8}{12} + \dfrac{1}{16x^6}$ from $x = 2$ to $x = 4$.

The length of this curve is $L = \displaystyle\int_2^4 \sqrt{1 + (f'(x))^2}\, dx$ where $f'(x) = $ [____].

Simplify and factor to get $L = \displaystyle\int_2^4 \sqrt{(g(x))^2}\, dx$ where $g(x) = $ [____].

Simplify and integrate to find $L = $ [____].

Section 7.4: Arc Length

8. A hawk flying at 18 m/s at an altitude of 120 m accidentally drops its prey. The parabolic trajectory of the falling prey is described by the equation $y = 120 - \dfrac{x^2}{54}$ until it hits the ground, where y is the height above the ground and x is the horizontal distance traveled in meters.

   Let D be the distance traveled by the prey from the time it is dropped until the time it hits the ground.

Work

Section 7.6

Before Class Video Examples

1. What are the work units of measurement?

2. If the force is constant, then Work = Force × Distance. Find the work done in lifting 15 lb. of books off the floor to the top of a table 4 ft high.

3. 3. A force of $3x^2 + 4x + 2$ pounds acts on a particle that is located a distance x from the origin. How much work is done in moving it from $x = 1$ to $x = 3$?

Section 7.6: Work

4. A 15-ft rope is lifted from the ground into the air by pulling it at constant a speed. The rope weighs 1.5 lb./ft. How much work was done lifting the rope 15 ft?

5. A 7–lb. bucket of water is lifted from the ground into the air by pulling 15 ft of rope at constant speed. The rope weighs 1.5 lb./ft. How much work was done lifting the bucket and rope?

Section 7.6: Work

Algebra Review

1. **Work**

Work, in physics, measure of energy transfer that occurs when an object is moved over a distance by an external force at least part of which is applied in the direction of the displacement.

Work = Force × Distance

2. **Equation of a Line**

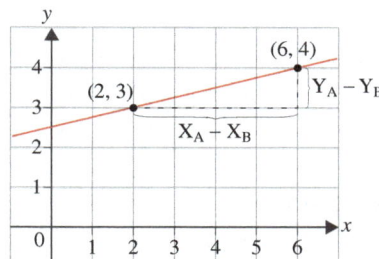

$y = mx + b$ where m = slope and b = y- intercept

$(y - y_0) = m \cdot (x - x_0)$ where m is the slope and (x_0, y_0) a point on the line

Example

i. Find the equation of the line that passes through points (3, −2) and (4, 5).

2. **Equation of a Circle**

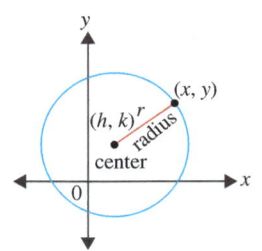

The **Standard Form** of the Equation of a Circle is $(x - h)^2 + (y - k)^2 = r^2$ where (h, k) is the center and r is the radius

Section 7.6: Work

Example

ii. Give the center and radius for $x^2 + y^2 = 49$

iii. Give the center and radius for $(x+1)^2 + (y-4)^2 = 25$

iv. Give the equation of a circle that has its center at the origin and radius 10.

v. Give the equation of a circle that has its center at the point $(0,5)$ and radius 2.

Work Done by Lifting

Work = Force × Distance

Force = Mass × Acceleration

Units:

	Force	Work
English	Pound (lb.)	Foot-pound (ft-lb.)
Metric	Newton (N)*	Newton-meter (N-m) or Joule (J)

*Amount in Newton = 9.8 · (Amount in kilogram)

Examples

1. How much work is done in lifting a 60-kg object 3 m up?

2. How much work is done in lifting a 60–lb. object 5 ft up?

3. Jessica McClure became famous at the age of 18 months after falling into a well in the backyard of her home in Texas on October 14, 1987. Between that day and October 16, rescuers worked for 58 hours to free "Baby Jessica" from the 8-inch wide well casing 22 feet below the ground. If a rescue worker pulled Jessica from the well using a rope with weight 3 lb./ft, and assuming that Jessica weighed 25 lb., how much work was done in pulling Jessica and the rope out of the well.

Section 7.6: Work

4. A 2-lb. bucket is lifted from the top of a 20-ft tall building by a cable at constant speed. The cable weighs 0.1 lb./ft.

   a. How much work is needed to lift the bucket and rope from the ground to the top of the building?

   b. How much work is needed to lift the bucket and rope from the ground half way up the building (from 0 ft to 10 ft)?

   c. How much work is needed to lift the bucket and rope the rest of the way to the top of the building (from 10 ft to 20 ft)?

C. Work Done by Springs

Hooke's Law: $F = kx$ $\quad F_{Spring} = k \cdot x$

$\downarrow \qquad \downarrow \qquad \downarrow$

Force $\quad$ Spring Constant $\quad$ Distance Stretched

Work: $W = \int_A^B kx \, dx$ where A = the point stretched from and B = the point stretched to.

Adjustments

Natural Length: Remember to make an adjustment for the natural length of a spring where appropriate.

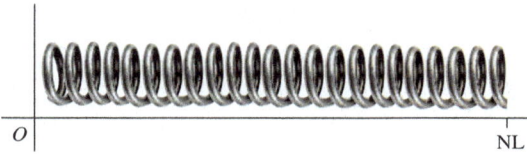

Units: Since the units for force and work are both based on lengths in feet or meters, we have to convert any other units (such as centimeter or inch) to feet or meter before starting any calculations.

Remember, $meter = 100 \, cm \implies x \, cm = \dfrac{x}{100} \, meter$, similarly, $ft = 12 \, in \implies x \, in = \dfrac{x}{12} \, ft$

Examples

1. (Type 1: **Force** given, **Work** asked) A force of 40 N is required to hold a spring that has been stretched from its natural position of 10 cm to a length of 15 cm. How much work is done in stretching the spring from 15 cm to 18 cm?

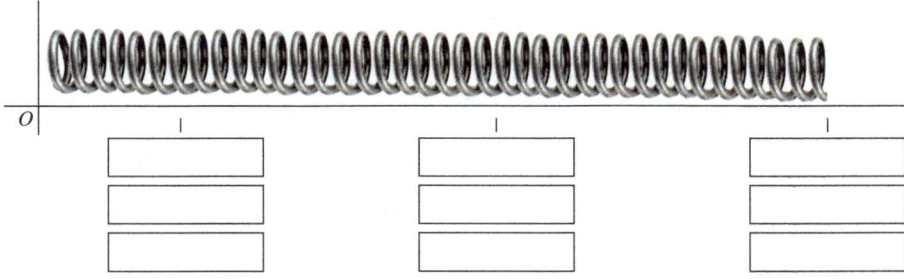

Section 7.6: Work

2. (Type 2: **Work** given, **Work/Force** asked) 7 Joules of work is required to stretch a spring from its natural length of 20 cm to 50 cm.

   a. How much work is done in stretching the spring from 25 cm to 35 cm?

   b. How far beyond its natural length will a force of 25 N hold the spring?

Section 7.6: Work

Work Done by Pumping

Water Density $= 1{,}000 \dfrac{kg}{m^3}$

Water Weight = Water Density × Gravity $= \left(1{,}000 \dfrac{kg}{m^3}\right) \cdot \left(9.8 \dfrac{m}{s^2}\right) = 9{,}800 \dfrac{kg}{m^2 \cdot s^2}$

Since Newton $= \dfrac{kg \cdot m}{s^2}$, Water Weight $= 9{,}800 \dfrac{N}{m^3}$

Water Weight:
- Metric Units: $9{,}800 \dfrac{N}{m^3}$
- English Units: $62.5 \dfrac{lb}{ft^3}$

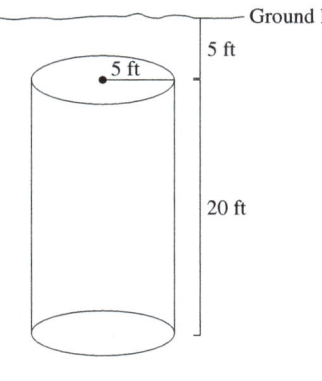

A cylindrical shaped tank is located 5 ft underground. If the tank is completely full of water, how much work is required to pump all of the water out of the tank to ground level?

Because of the properties of water pressure, the distance that the water is being pumped (lifted) will be the distance from the surface of the water to the location it is pumped to. However, during the process of pumping, the water level will go down, and hence the distance will continuously change.

So, to find the total amount of work done, we will find the work done in pumping one "slice" of water out of the tank, and then add up all of the slices.

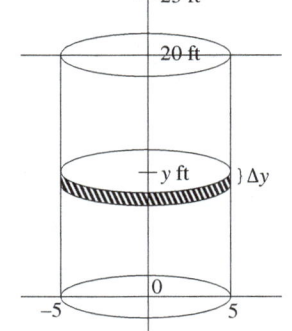

Consider a slice made at a random height y.

The work done in pumping out this slice can be calculated by:

W_{SLICE} = Force · Distance

And we know that Force = Volume of the Slice · Water Weight, so

W_{SLICE} = Volume · Water Weight · Distance

Since the slice has a cylindrical shape,
- Volume of the Slice =

- Water Weight =

- Distance water is lifted =

Section 7.6: Work

W_{SLICE} = Volume · Water Weight · Distance

W_{TOTAL} =

Examples

1. An inverted cone shaped tank has dimensions: height C = 12 m, radius A = 4 m. If the tank is completely full, find the work done in pumping all the water out of the tank to a level of B = 3 m above the top of the tank.

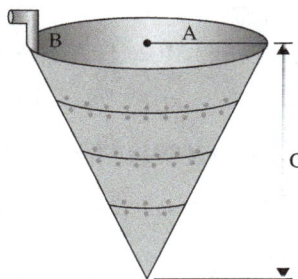

Section 7.6: Work

Rule for Pumping

- General Tank: $W = \int_A^B (\text{Area of Slice}) \cdot (\text{Water Weight}) \cdot (C - y)\, dy$

- Tank with a **circular** cross section: $W = \int_A^B \pi (\text{Radius of Tank})^2 \cdot (\text{Water Weight}) \cdot (C - y)\, dy$

where

- A is the "bottom level" of water to be removed
- B is the "top level" of water to be removed
- C is the level that water is pumped to

2. A trough shaped tank has dimensions: a = 4 ft, b = 8 ft, h = 12 ft, and = 20 ft. If the tank is completely full, find the work done in pumping all the water out of the tank, over the side of the tank.

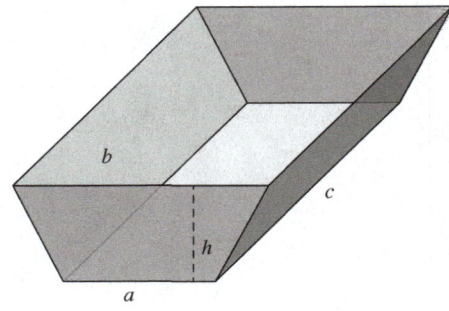

3. A hemispherical shaped tank has dimensions: r = 10 m. If the tank is completely full, find the work done in pumping all the water out of the tank, to a distance of 2 m above the top of the tank.

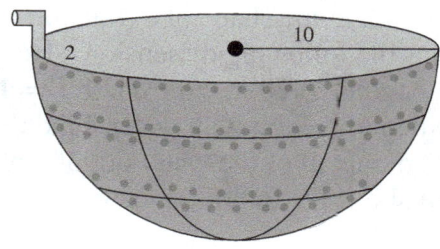

Section 7.6: Work

4. A tank in the shape of a triangular prism (see figure) has the following dimensions: Height **B = 15 m**, Length **C = 20 m**, and Width **D = 10 m**. The tank is filled to the top with water. Find the work required to pump all of the water out of tank to a height **A = 5 m** above the top of the tank.

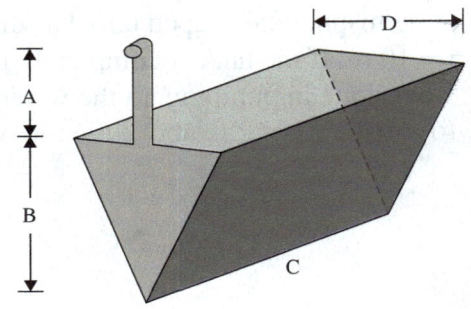

5. A tank in the shape of a cylinder on its side (see figure) has the following dimensions: Height **H = 15 m** and Radius **R = 5 m**. The tank is filled to the top with water. Find the work required to pump all of the water out of a hole in the top of the tank.

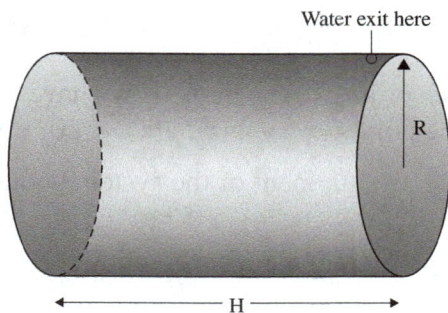

Section 7.6: Work

E. Before Class Video Examples (Center of Mass)

1. According to the book, if we have a system of n particles with masses $m_1, m_2, \ldots m_n$ located at the points $(x_1, y_1), (x_2, y_2), \ldots (x_n, y_n)$, in the xy-plane, then:
   - The moment of the system about the y-axis is $M_y = \sum_{i=1}^{n} m_i x_i$
   - The moment of the system about the x-axis is $M_x = \sum_{i=1}^{n} m_i y_i$

2. Consider the three masses $m_1 = 2$, $m_2 = 1$, $m_3 = 3$, which are located at the points $P_1 = (0, 4)$, $P_2 = (7, 3)$, and $P_3 = (3, 7)$. Find M_x, the moment of the system about the x-axis.

3. Consider the three masses $m_1 = 2$, $m_2 = 1$, $m_3 = 3$, which are located at the points $P_1 = (0, 4)$, $P_2 = (7, 3)$, and $P_3 = (3, 7)$. Find M_y, the moment of the system about the y-axis.

4. 4. Consider the three masses $m_1 = 2$, $m_2 = 1$, $m_3 = 3$, which are located at the points $P_1 = (0, 4)$, $P_2 = (7, 3)$, and $P_3 = (3, 7)$. Find the coordinates of the center of mass (centroid) of the system.

Section 7.6: Work

5. According to the book, If the region R lies between two curves $y = f(x)$ and $y = g(x)$, where $f(x) \geq g(x)$, then $\bar{x} = \dfrac{1}{A}\int_a^b x[f(x) - g(x)]dx$ and $\bar{y} = \dfrac{1}{A}\int_a^b \dfrac{1}{2}\left[(f(x))^2 - (g(x))^2\right]dx$.

6. Find the coordinates $(\bar{x}, \bar{y})$ of the centroid of the region bounded by $y = x^2$ and $y = 8\sqrt{x}$ on the interval [0, 4].

F. Center of Mass

	Discrete	Continuous $f(x)$ = Top and $g(x)$ = Bottom
Moment about the y-axis	$M_y = \sum_{i=1}^{n} m_i x_i$	$M_y = \rho \int_a^b x \cdot [f(x) - g(x)] dx$
Moment about the x-axis	$M_x = \sum_{i=1}^{n} m_i y_i$	$M_x = \frac{\rho}{2} \int_a^b f(x)^2 - g(x)^2 \, dx$
Mass	$\text{Mass} = \sum_{i=1}^{n} m_i$	$\text{Mass} = \rho \int_a^b f(x) - g(x) \, dx$
C.O.M.	$(\bar{x}, \bar{y}) = \left(\dfrac{M_y}{\text{Mass}}, \dfrac{M_x}{\text{Mass}} \right)$	$(\bar{x}, \bar{y}) = \left(\dfrac{M_y}{\text{Mass}}, \dfrac{M_x}{\text{Mass}} \right)$

Examples

1. Find M_x, M_y, and the center of mass of the system of objects with masses

    (−2, 1) mass 2

    (1, −1) mass 5

    (4, 3) mass 3

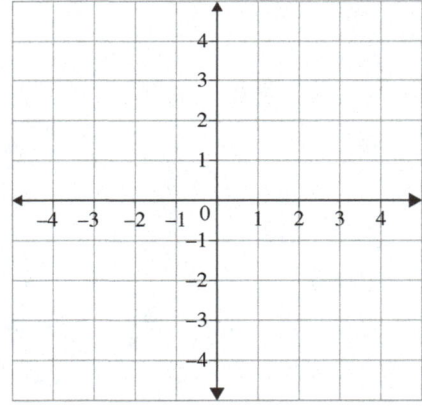

2. Find the centroid lying between the regions $f(x) = \sqrt{x}$ and $g(x) = x^2$

Section 7.6: Work

Function of y given

	Discrete	Continuous $f(y)=$ Right and $g(y)=$ Left
Moment about the y-axis	$M_y = \sum_{i=1}^{n} m_i x_i$	$M_y = \dfrac{\rho}{2}\int_a^b f(y)^2 - g(y)^2 \, dy$
Moment about the x-axis	$M_x = \sum_{i=1}^{n} m_i y_i$	$M_x = \rho \int_a^b y \cdot [f(y) - g(y)] \, dy$
Mass	$\text{Mass} = \sum_{i=1}^{n} m_i$	$\text{Mass} = \rho \int_a^b f(y) - g(y) \, dy$
C.O.M.	$(\bar{x}, \bar{y}) = \left(\dfrac{M_y}{\text{Mass}}, \dfrac{M_x}{\text{Mass}} \right)$	$(\bar{x}, \bar{y}) = \left(\dfrac{M_y}{\text{Mass}}, \dfrac{M_x}{\text{Mass}} \right)$

WeBWorK

Section 7.6 Work:

1. A particle is moved along the x-axis by a force that measures $2x^3 + 5$ pounds at a point x feet from the origin. Find the work done in moving the particle from the origin to a distance of 2 ft.

Section 7.5 Centers:

6. A lamina has the shape of a triangle with vertices at (−10, 0), (10, 0), and (0, 8). Its density is $\rho = 4$.

   A. What is the total mass?

   B. What is the moment about the x-axis?

   C. What is the moment about the y-axis?

   D. Where is the center of mass?

Sequence

Section 8.1

Before Class Video Examples

1. Find the first four elements of the sequence $\dfrac{n+1}{2n-1}$

2. Find the formula for the nth term of the sequence $1, -\dfrac{1}{8}, \dfrac{1}{27}, -\dfrac{1}{64}, \dfrac{1}{125}\ldots$

Section 8.1: Sequence

3. Find the limit of the sequence if it converges, otherwise state divergent.

   a. $\left\{\dfrac{3n+2}{4n+5}\right\}$

   b. $\left\{\dfrac{3n^2+2}{4n+5}\right\}$

   c. $\left\{\dfrac{3n+2}{4n^2+5}\right\}$

   d. $\left\{\dfrac{3+(-1)^n}{3}\right\}$

Algebra Calculus Review

1. **Functions**

> DEFINITION: **Function**—A correspondence from one set (the domain) to another set (the range) such that each element in the domain corresponds to exactly one element in the range.
>
> DEFINITION: **Domain**—Input → x-values
>
> DEFINITION: **Range**—Output → y-values

2. Limit Laws

1. $\lim\limits_{n\to\infty}(a_n + b_n) = \lim\limits_{n\to\infty} a_n + \lim\limits_{n\to\infty} b_n$

2. $\lim\limits_{n\to\infty}(a_n \cdot b_n) = \lim\limits_{n\to\infty} a_n \cdot \lim\limits_{n\to\infty} b_n$

3. $\lim\limits_{n\to\infty}\left(\dfrac{a_n}{b_n}\right) = \dfrac{\lim\limits_{n\to\infty} a_n}{\lim\limits_{n\to\infty} b_n}; \quad \lim\limits_{n\to\infty} b_n \neq 0$

4. $\lim\limits_{n\to\infty} C \cdot a_n = C \cdot \lim\limits_{n\to\infty} a_n$

5. $\lim\limits_{n\to\infty}(a_n)^P = \left(\lim\limits_{n\to\infty} a_n\right)^P; \quad if\ p > 0\ and\ a_n > 0$

6. $\lim\limits_{n\to\infty} n = \infty$

7. $\lim\limits_{n\to\infty} \dfrac{1}{n} = 0$

8. $\lim\limits_{n\to\infty} \dfrac{1}{n^C} = 0 \text{ for } C \geq 1$

9. $\lim\limits_{n\to\infty} a^n = \infty \text{ for } a \geq 1$

A. Sequence

Examples: Give the sequence for each of the following:

1. 2, 4, 6, 8, . . .

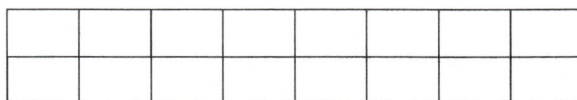

2. $\dfrac{1}{5}, \dfrac{1}{25}, \dfrac{1}{125}\ldots$

Section 8.1: Sequence

3. 3, 5, 7, . . .

4. $\dfrac{1}{2}, \dfrac{3}{5}, \dfrac{5}{8}, \dfrac{7}{11}...$

B. Function versus Sequence

	Function $f(x)$	Sequence a_n
Domain:	$x \in \mathbb{R}$ i.e., the domain can be any value of x that is a real number	$n \in \mathbb{N}$ i.e., the domain can be any value of n that is a natural number (non-negative integer)
Range:	Continuous	Discreet
Graph (Example)	$f(x) = 2x$	$a_n = 2n$

Section 8.1: Sequence

C. Limit of the Sequence—Converge or Diverge?

$$\lim_{n \to \infty} a_n = C \to \text{Converge}$$

$$\lim_{n \to \infty} a_n = \pm\infty \to \text{Diverge}$$

Examples: Give the limit for each of the following sequences:

5. $\lim\limits_{n \to \infty} 2n =$

6. $\lim\limits_{n \to \infty} \dfrac{1}{5^n} =$

7. $\lim\limits_{n \to \infty} \dfrac{2n-1}{3n-1} =$

D. Review of Limit Laws

Examples: Give the limit for each of the following sequences:

8. $a_n = \dfrac{3}{\sqrt{n}} - \dfrac{4n^2+1}{7-8n^2}$

Section 8.1: Sequence

9. $b_n = \dfrac{3n}{4n+1} \cdot \tan\left(\dfrac{5\pi n + 1}{20n+3}\right)$

10. $d_n = (-1)^n$

11. $a_n = \dfrac{(-1)^n}{n+3}$

12. $k_n = \dfrac{3^{-n}}{n^2 + 4}$

E. Monotonic Sequences

A Sequence a_n is monotonically **increasing** if $a_n \leq a_{n+1}$ for all n.

A Sequence a_n is monotonically **decreasing** if $a_n \geq a_{n+1}$ for all n.

F. Bounded Sequences

A Sequence a_n is bounded from **above** if $a_n \leq M$ (we call M the Upper Bound)

A Sequence a_n is bounded from **below** if $a_n \geq M$ (we call M the Lower Bound)

for some finite number M and for all n.

Theorem
- Every sequence that is monotonically **decreasing** and is bounded from **below** is convergent.
- Every sequence that is monotonically **increasing** and is bounded from **above** is convergent.

Example: Does the sequence converge or diverge?

13. $\left\{\dfrac{2}{n}\right\}$

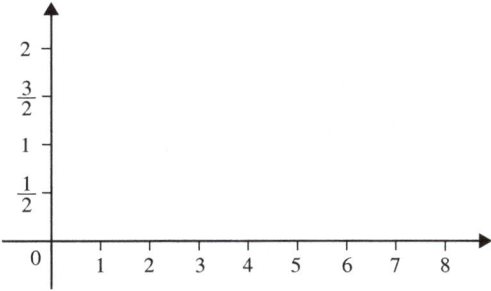

Section 8.1: Sequence

G. Factorial (!)

> Definition: $a! = a \cdot (a-1) \cdot (a-2) \cdots 3 \cdot 2 \cdot 1$

Example: Does the sequence converge or diverge?

14. $5!$

15. $\dfrac{8!}{4!}$

16. $\dfrac{(n+2)!}{n!}$

WeBWorK

5. Determine whether the sequence is divergent or convergent. If it is convergent, evaluate its limit. If it diverges to infinity, state your answer as "INF" (without the quotation marks). If it diverges to negative infinity, state your answer as "MINF." If it diverges without being infinity or negative infinity, state your answer as "DIV."

The sequence $a_n = \dfrac{(n+3)!}{n!}$, then $\lim\limits_{n \to \infty} a_n =$

The sequence $b_n = \dfrac{n!}{(n+3)!}$, then $\lim\limits_{n \to \infty} b_n =$

14. Determine whether the sequences are increasing, decreasing, or not monotonic. If increasing, enter 1 as your answer. If decreasing, enter −1 as your answer. If not monotonic, enter 0 as your answer.

   i. $a_n = \dfrac{n-3}{n+3}$

   ii. $a_n = \dfrac{\cos n}{3^n}$

   iii. $a_n = \dfrac{1}{3n+9}$

15. Write down the first five terms of the following recursively defined sequence. $a_1 = 3$, $a_{n+1} = 4 - \dfrac{1}{a_n}$

$a_2 =$

$a_3 =$

$a_4 =$

$a_5 =$

Then $\lim_{n\to\infty} a_n =$

Series

Section 8.2

Before Class Video Examples

1. Determine the first term of the series, the common ratio r and determine whether each converges or diverges.

   a. $\sum_{n=1}^{\infty} \dfrac{3}{8^n}$

   b. $\sum_{n=1}^{\infty} \dfrac{4^n}{3^{n+1}}$

2. Determine whether the series converges or diverges.

   $\sum_{n=1}^{\infty} \dfrac{14}{9n}$

Section 8.2: Series

3. Determine whether the series converges or diverges.

$$\sum_{n=1}^{\infty} \frac{n^2+4}{5n^2+7}$$

Algebra Review

1. Exponents

Exponential Notation

$b^n = \underbrace{b \cdot b \cdot b \cdot \ldots \cdot b}_{n \text{ times}},$ Example: $3^4 = 3 \cdot 3 \cdot 3 \cdot 3 = 81$

Common Rules of Exponents

- $x^a \cdot x^b = x^{(a+b)}$ (Product Rule)
- $\dfrac{x^a}{x^b} = x^{(a-b)}$ (Quotient Rule)
- $x^0 = 1$ (Zero-Exponent Rule)
- $\dfrac{1}{x^b} = x^{(-b)}$ or $\dfrac{1}{x^{-b}} = x^b$ (Negative Exponent Rule)

- $(x^a)^b = x^{a \cdot b}$ (Power Rule)
- $(x \cdot y)^a = x^a \cdot y^a$ (Products to Power)
- $\left(\dfrac{x}{y}\right)^a = \dfrac{x^a}{y^a}$ (Quotients to Power)

Example

i. $a^3 \cdot a^6 = a^{\square}$

ii. $x^{10} = x^7 \cdot x^{\square}$

iii. $5 \cdot 2^n \cdot 8^n =$

iv. Express 4^{n+3} in the form $a \cdot r^{n-1}$

v. Express $\dfrac{7^{n+1}}{4^n}$ in the form $a \cdot r^{n-1}$

vi. Express $\dfrac{3^{3n}}{4^{n-3}}$ in the form $a \cdot r^{n-1}$

A. Series

> Definition: A series is the sum of a sequence.
> For a sequence $a_n = \{a_1, a_2, a_3, a_4, a_5, ...\}$
> we have the corresponding series $S_n = \sum\limits_{n=1}^{\infty} a_n = a_1 + a_2 + a_3 + a_4 + a_5 + \cdots$

Example
1. Write out the sequence and series, and whether each will converge or diverge for $\left\{3\left(\dfrac{1}{10}\right)^n\right\}$.

Section 8.2: Series

B. Partial Sums

$$S_1 = \sum_{n=1}^{1} a_n = a_1$$

$$S_2 = \sum_{n=1}^{2} a_n = a_1 + a_2$$

$$S_k = \sum_{n=1}^{k} a_n = a_1 + a_2 + a_3 + a_4 + a_5 + \cdots + a_k$$

Example

2. For the sequence $\left\{3\left(\dfrac{1}{10}\right)^n\right\}$ find S_1, S_2, S_3, and S_k.

C. The Geometric Series

Theorem

$$\sum_{n=1}^{\infty} ar^{n-1} = \sum_{n=0}^{\infty} ar^n \Rightarrow \begin{cases} \text{converge to } \dfrac{a}{1-r} & \text{if } |r|<1 \\ \text{diverge} & \text{if } |r|\geq 1 \end{cases}$$

Goal: Tell whether a series converges or diverges. If it converges, what to?

Examples: Determine whether each series will converge or diverge. If convergent, what to?

3. $\sum_{n=1}^{\infty} 3\left(\dfrac{1}{10}\right)^{n-1}$

4. $\sum_{n=1}^{\infty} 4(-1)^{n+1}\left(\dfrac{3}{4}\right)^{n+1}$

Section 8.2: Series

5. $\displaystyle\sum_{n=1}^{\infty} 2^{2n}\, 3^{1-n}$

6. $\displaystyle\sum_{n=3}^{\infty} -2\left(\frac{1}{3}\right)^{n+1}$

D. The Telescoping Series

Telescoping Series is a technique used when summing a sequence that is in the form of a fraction; usually has the same numerator and has n in the denominator.

Examples: Determine whether each series will converge or diverge. If convergent, what to?

7. $\sum_{n=2}^{\infty} \dfrac{3}{n-1} - \dfrac{3}{n+1}$

Section 8.2: Series

8. $\displaystyle\sum_{n=5}^{\infty} \frac{3}{n^2-n-6}$

E. The Divergence Test

Theorem: If a_n does not converge to 0 (i.e., $\lim\limits_{n\to\infty} a_n \neq 0$) then $\sum\limits_{n=1}^{\infty} a_n$ will diverge.

Examples: Determine whether each series will converge or diverge.

9. $\sum\limits_{n=1}^{\infty} \dfrac{n^2 + 4}{5n^2 + n + 7}$

*Note: If $\lim\limits_{n\to\infty} a_n = 0$, the Theorem is inconclusive. It does not necessarily imply that the series will converge!!!

F. The Harmonic Series

Theorem: $\sum\limits_{n=1}^{\infty} \dfrac{1}{n}$ diverges.

Examples: Determine whether each series will converge or diverge.

10. $\sum\limits_{n=1}^{\infty} \dfrac{3}{5n}$

G. Combinations of Series

If $\sum\limits_{n=1}^{\infty} a_n$ and $\sum\limits_{n=1}^{\infty} b_n$ are both convergent series, then $\sum\limits_{n=1}^{\infty} (a_n + b_n)$ and $\sum\limits_{n=1}^{\infty} C \cdot a_n$ are also convergent series.

Section 8.2: Series

WeBWorK

7. Decide whether the given sequence or series is convergent or divergent. If convergent, enter the limit (for a sequence) or the sum (for a series). If divergent, enter DIV.

   a. The sequence $\left\{\dfrac{5}{4n}\right\}$

   b. The series $\sum_{n=1}^{\infty} \dfrac{5}{4n}$

8. Determine the sum of the following series:
$$\sum_{n=1}^{\infty} \dfrac{2^n + 5^n}{8^n}$$

11. Determine whether the following series is convergent or divergent. $9 - \dfrac{9}{5} + \dfrac{9}{25} - \dfrac{9}{125} + \cdots =$

14. Express 2.927927 as a rational number, in the form $\dfrac{p}{q}$ where p and q are positive integers with no common factors.

15. For the series $\sum_{n=1}^{\infty} \dfrac{x^n}{4^n}$

   a. Find the values of x for which the series converges.

   b. Find the sum of the series for those values of x.

Integral and Comparison Tests

Section 8.3

Before Class Video Examples

1. Determine whether the series converges or diverges.

   a. $\sum_{n=1}^{\infty} 3ne^{-n^2}$

2. Determine whether the series converges or diverges.

   a. $\sum_{n=1}^{\infty} \dfrac{1}{\sqrt[3]{n}}$

   b. $\sum_{n=1}^{\infty} \dfrac{1}{n^4}$

Section 8.3: Integral and Comparison Tests

3. Determine whether the series converges or diverges.

   a. $\displaystyle\sum_{n=1}^{\infty} \frac{1}{n-3}$

   b. $\displaystyle\sum_{n=1}^{\infty} \frac{1}{n^2-3}$

Algebra Review

1. **Function Inequalities**

   A function f is greater than or equal to g if $f(x) \geq g(x)$ for all $x \in$ Domain

   Similarly, a sequence a_n is greater than or equal to b_n if $a_n \geq b_n$ for all $n \in$ Domain

   Example
   i. $5n^2 + 12$ ☐ $5n^2 + 1$

   ii. $\dfrac{1}{5n^2 + 12}$ ☐ $\dfrac{1}{5n^2 + 1}$

Section 8.3: Integral and Comparison Tests

A. Integral Test

Let $a_n = f(n)$, where $f(x)$ is *positive, decreasing,* and *continuous* for $x \geq 1$. Then,

i. If $\int_1^\infty f(x)\,dx$ converges $\Rightarrow \sum_{n=1}^\infty a_n$ converges.

ii. If $\int_1^\infty f(x)\,dx$ diverges $\Rightarrow \sum_{n=1}^\infty a_n$ diverges.

Examples: Determine whether each of the following series is convergent or divergent.

1. $\sum_{n=2}^\infty \dfrac{4}{n \cdot (\ln(n))^3}$

2. $\sum_{n=1}^{\infty} \dfrac{1}{n}$

3. $\sum_{n=1}^{\infty} \dfrac{1}{n^2}$

B. P-Series

$$\sum_{n=1}^{\infty} \frac{1}{n^P} = \begin{cases} \text{Convergent if } P > 1 \\ \text{Divergent if } P \leq 1 \end{cases}$$

Examples: Determine whether each of the following series is convergent or divergent.

4. $\sum_{n=4}^{\infty} \frac{12}{n^3}$

5. $\sum_{n=1}^{\infty} 108 n^4$

C. Direct Comparison Test

For two sequences a_n and b_n such that $0 \leq a_n \leq b_n$ (i.e., both sequences are positive),

i. If $\sum_{n=1}^{\infty} b_n$ converges $\Rightarrow \sum_{n=1}^{\infty} a_n$ also converges.

ii. If $\sum_{n=1}^{\infty} a_n$ diverges $\Rightarrow \sum_{n=1}^{\infty} b_n$ also diverges.

Examples: Determine whether each of the following series is convergent or divergent.

3. $\sum_{n=1}^{\infty} \frac{n^2 - 1}{n^5 + 5}$

4. $\sum_{n=1}^{\infty} \dfrac{5}{n^2+4}$

5. $\sum_{n=1}^{\infty} \dfrac{3}{4n-1}$

6. $\sum_{n=1}^{\infty} \dfrac{3}{2^{n-1}+5}$

D. Limit Comparison Test

For $\sum_{n=1}^{\infty} a_n$ and $\sum_{n=1}^{\infty} b_n$ two positive term series;

If $\lim_{n \to \infty} \dfrac{a_n}{b_n} = C$ (where $C > 0$ and C is finite), then both series will either converge or both will diverge.

Examples: Determine whether each of the following series is convergent or divergent.

7. $\sum_{n=1}^{\infty} \dfrac{n^2 + 2n - 5}{n^4 + n + 7}$

8. $\sum_{n=1}^{\infty} \sin\left(\dfrac{1}{n}\right)$

9. $\displaystyle\sum_{n=2}^{\infty} \frac{1}{n^3 - n}$

10. $\displaystyle\sum_{n=1}^{\infty} \frac{\sin^2 n}{n \cdot \sqrt{n}}$

Section 8.3: Integral and Comparison Tests

WeBWorK

3. Each of the following statements is an attempt to show that a given series is convergent or divergent. For each statement, enter C (for "correct") if the argument is valid, or enter I (for "incorrect") if any part of the argument is flawed. (Note: if the conclusion is true but the argument that led to it was wrong, you must enter I.)

[ ] 1. For all $n > 2$, $\dfrac{1}{n \cdot \ln n} < \dfrac{1}{n}$, and the series $\sum \dfrac{1}{n}$ diverges.

So by the Comparison Test, the series $\sum \dfrac{1}{n \cdot \ln n}$ diverges.

[ ] 2. For all $n > 3$, $\dfrac{1}{n^2 - 4} < \dfrac{1}{n^2}$, and the series $\sum \dfrac{1}{n^2}$ converges.

So by the Comparison Test, the series $\sum \dfrac{1}{n^2 - 4}$ converges.

[ ] 3. For all $n > 1$, $\dfrac{1}{n^2 + n + 4} < \dfrac{1}{n^2}$, and the series $\sum \dfrac{1}{n^2}$ converges.

So by the Comparison Test, the series $\sum \dfrac{1}{n^2 + n + 4}$ converges.

7. Select the FIRST correct reason why the given series diverges.

   A. Divergent p-series
   B. Divergent geometric series
   C. Comparison with a divergent p-series
   D. Diverges because the terms don't have limit zero
   E. Integral test

   ☐ 1. $\sum_{n=3}^{\infty} \dfrac{1}{n \cdot \ln n}$

   ☐ 2. $\sum_{n=3}^{\infty} \dfrac{\ln n}{n}$

   ☐ 3. $\sum_{n=3}^{\infty} \ln n$

   ☐ 4. $\sum_{n=3}^{\infty} \dfrac{1}{n}$

Alternating Series Test

Section 8.4

Before Class Video Examples

1. Determine whether the series converges or diverges.

   a. $\sum_{n=1}^{\infty} \frac{(-1)^{n+1}}{n}$

   b. $\sum_{n=1}^{\infty} \frac{(-1)^n n}{2n-1}$

Section 8.4: Alternating Series Test

2. Determine whether the series is absolutely convergent, conditionally convergent, or divergent.

   a. $\displaystyle\sum_{n=1}^{\infty} \frac{(-1)^n}{3+n^4}$

   b. $\displaystyle\sum_{n=1}^{\infty} \frac{(-1)^{n+1}}{\sqrt[3]{n}}$

   c. $\displaystyle\sum_{n=1}^{\infty} \frac{(-1)^n\, n}{3+n}$

Section 8.4: Alternating Series Test

3. Determine whether the series converges or diverges.

   a. $\displaystyle\sum_{n=1}^{\infty} \frac{3^n}{4n!}$

   b. $\displaystyle\sum_{n=1}^{\infty} \frac{(-3)^n}{n}$

Section 8.4: Alternating Series Test

Algebra/Calculus Review

1. **Determining Slope**

   Increasing/Decreasing Test
   - If $f'(x) > 0$ on an interval, then $f(x)$ is increasing on that interval
   - If $f'(x) < 0$ on an interval, then $f(x)$ is decreasing on that interval

   Example

   i. Find the interval of increase for the function $f(x) = 3x^2 + 18x$.

   ii. Does the sequence $a_n = \dfrac{3n+3}{2n^2}$ increase or decrease on the interval $(0, \infty)$ $a_n = \dfrac{3n+3}{2n^2}$?

A. Alternating Series

Alternating Series: $a_n = (-1)^n$
$b_n = \cos(n\pi)$

$$\sum_{n=1}^{\infty}(-1)^{n+1} \cdot b_n = b_1 - b_2 + b_3 - b_4 + \cdots$$

    i. If $b_{n+1} \leq b_n$ for all n (i.e., sequence is decreasing)

    ii. If $\lim\limits_{n \to \infty} b_n = 0$

Then $\sum\limits_{n=1}^{\infty}(-1)^{n+1} \cdot b_n$ converges. If not, it diverges.

Examples: Determine whether each of the following series is convergent or divergent.

1. $\sum\limits_{n=1}^{\infty} \dfrac{(-1)^{n+1}}{n}$

2. $\sum\limits_{n=1}^{\infty} \dfrac{(-1)^n \cdot 3n}{4n-1}$

3. $\sum\limits_{n=1}^{\infty} \dfrac{\cos(n\pi) \cdot n}{n^3 + 1}$

B. The Alternating Series Estimation Theorem

If is the sum $S = \sum_{n=1}^{\infty}(-1)^{n-1}\cdot b_n$ of the alternating series that satisfies

i. If $b_{n+1} \leq b_n$ for all n (i.e., sequence is decreasing)

ii. If $\lim_{n \to \infty} b_n = 0$

$$\underbrace{|R_n|}_{\substack{\text{Remainder} \\ \text{after } n \text{ terms}}} = \underbrace{|S}_{\substack{\text{True} \\ \text{Sum}}} - \underbrace{S_n|}_{\substack{\text{Partial} \\ \text{Sum}}} \leq b_{n+1}$$

Examples

4. $\sum_{n=1}^{\infty} \dfrac{(-1)^n}{n}$

   a. Determine whether the series is convergent or divergent.

   b. Find the partial sums, S_5 through S_8.

   c. Give an upper bound for R_5 (i.e., give the maximum value for the difference between R_5 and the true sum).

   d. For the difference between S_n and the true sum to be within 0.000001, to which term should S_n be calculated (i.e., what is the smallest value of n to guarantee the error less than 0.000001)?

5. $\sum_{n=1}^{\infty} \dfrac{(-1)^n \cdot 2}{(3)^n}$

   a. Determine whether the series is convergent or divergent.

   b. Find the partial sum S_4.

   c. Give the true sum S_n if possible.

   d. Find the true difference between S_n and the partial sum S_4. Compare the value to R_4 calculated by the estimation formula.

   e. For the difference between S_n and the true sum to be within 0.001, to which term should S_n be calculated? (i.e., what is the smallest value of n to guarantee the error less than 0.001)?

Section 8.4: Alternating Series Test

6. $\sum_{n=1}^{\infty} \dfrac{(-1)^{n+1}}{n!}$

    a. Determine whether the series is convergent or divergent.

    b. Find the partial sum S_4.

    c. Find R_4.

    d. What should n be for R_n to be accurate within 0.0001?

C. Absolute Convergence

Alternating Series may:

- Diverge
- Conditionally converge
- Absolutely converge

> If $\sum_{n=1}^{\infty} |a_n|$ is convergent, then $\sum_{n=1}^{\infty} (-1)^n \cdot a_n$ is ABSOLUTELEY convergent.

- If you have shown that $\sum_{n=1}^{\infty} (-1)^n \cdot a_n$ is divergent, you are done.

- If you have shown that $\sum_{n=1}^{\infty} (-1)^n \cdot a_n$ is convergent, it remains to be shown that the convergence is absolute or conditional.
    - If you sequentially show that $\sum_{n=1}^{\infty} |a_n|$ is convergent, you have ABSOLUTE convergence.
    - If you sequentially show that $\sum_{n=1}^{\infty} |a_n|$ is divergent, you have CONDITIONAL convergence.

Alternatively

- If, in your first step, you are able to show that $\sum_{n=1}^{\infty} |a_n|$ is convergent, then $\sum_{n=1}^{\infty} (-1)^n \cdot a_n$ is ABSOLUTELY convergent.

- If, in your first step, you show that $\sum_{n=1}^{\infty} |a_n|$ is divergent, $\sum_{n=1}^{\infty} (-1)^n \cdot a_n$ might still be CONDITIONALY convergent or divergent.

Examples

7. $\sum_{n=1}^{\infty} \frac{(-1)^{n+1}}{n^3}$

8. $\sum_{n=1}^{\infty} \frac{(-1)^n}{n}$

D. Ratio Test

For $\sum_{n=1}^{\infty} a_n$ (a general or alternating series) If $\lim_{n \to \infty} \left| \dfrac{a_{n+1}}{a_n} \right| = L$, and

- $L < 1 \Rightarrow \sum_{n=1}^{\infty} a_n$ will Absolutely Converge
- $L > 1 \Rightarrow \sum_{n=1}^{\infty} a_n$ will Diverge
- $L = 1 \Rightarrow \sum_{n=1}^{\infty} a_n$ Inconclusive (Try another test)

*Use this test whenever the sequence contains a factorial.

* Be careful not to confuse this with the Limit Comparison Test!

Examples

9. $\sum_{n=1}^{\infty} \dfrac{3^n}{4n!}$

10. $\sum_{n=1}^{\infty} \dfrac{1}{n^3}$

11. $\sum_{n=1}^{\infty} \dfrac{(-3)^n}{n^3}$

General Examples (Determine which test to use).

12. $\sum_{n=1}^{\infty} (-1)^n \dfrac{n}{n^2+1}$

Section 8.4: Alternating Series Test

13. $\sum_{n=1}^{\infty} \dfrac{10^n}{(n+1) \cdot 4^{2n+1}}$

14. $\sum_{n=1}^{\infty} 3^n \cdot 4^{1-n}$

E. Root Test

For $\sum_{n=1}^{\infty} a_n$ (a general or alternating series) If $\lim_{n \to \infty} \sqrt[n]{|a_n|} = L$, and

- $L < 1 \Rightarrow \sum_{n=1}^{\infty} a_n$ will Absolutely Converge
- $L > 1 \Rightarrow \sum_{n=1}^{\infty} a_n$ will Diverge
- $L = 1 \Rightarrow \sum_{n=1}^{\infty} a_n$ Inconclusive (Try another test)

Examples

15. $\sum_{n=1}^{\infty} \left(\dfrac{n^2 + 8}{2n^2 - 3n} \right)^n$

16. $\sum_{n=1}^{\infty} \dfrac{(-n)^n}{(\ln n)^n}$

Section 8.4: Alternating Series Test

F. Calculator

17. Use your calculator to evaluate $\sum_{n=1}^{\infty} 3^n \cdot 4^{1-n}$.

TI 83/84 (Old):

[Second] [START]   (List)

→ Scroll to "**Math**" – Choose **5: sum(**

[Second] [START]

→ Scroll to "**OPS**"—Choose **5: seq(**

→ Enter the *fxn* , *x* , *lowerbound* , *upperbound*)

(For infinity, enter 100)

[ENTER]

TI 84 (New):

[Second] [START]   (List)

→ Scroll to "**Math**" – Choose **5: sum(**

[Second] [START]

→ Scroll to "**OPS**"—Choose **5: seq(**

→ Enter the Expr: *fxn*

　　　　Variable: *x*

　　　　start: *lowerbound*

　　　　end: *upperbound*

　　　　step: **1**

(For infinity, enter 100)

→ Click on "Paste"

[ENTER]

✦ WeBWorK

1. Match each of the following series with the correct statement:

   A. The series is absolutely convergent.

   C. The series is conditionally convergent.

   D. The series diverges.

$$\sum_{n=1}^{\infty}(-1)^n \frac{7n}{2n-1}$$

$$\sum_{n=1}^{\infty}\frac{(-1)^{n-1}}{n^4}$$

$$\sum_{n=1}^{\infty}\frac{(-1)^n}{n\sqrt{n}}$$

$$\sum_{n=1}^{\infty}\frac{(-1)^{n-1}}{n}$$

Section 8.4: Alternating Series Test

For # 6 and 8: Consider the series. Attempt the Ratio Test to determine whether the series converges. Give $\left|\dfrac{a_{n+1}}{a_n}\right| = \boxed{}$, $L = \lim\limits_{n\to\infty}\left|\dfrac{a_{n+1}}{a_n}\right| = \boxed{}$

Which of the following statements is true?

A. The Ratio Test says that the series converges absolutely.
B. The Ratio Test says that the series diverges.
C. The Ratio Test says that the series converges conditionally.
D. The Ratio Test is inconclusive, but the series converges absolutely by another test or tests.
E. The Ratio Test is inconclusive, but the series diverges by another test or tests.
F. The Ratio Test is inconclusive, but the series converges conditionally by another test or tests.

6. $\displaystyle\sum_{n=1}^{\infty}(-1)^n\dfrac{n^8}{5^n}$

8. $\displaystyle\sum_{n=1}^{\infty}(-1)^n\dfrac{n}{n^2+6}$

Section 8.4: Alternating Series Test

11. Match each of the following series with the first correct statement.

   A. The series is absolutely convergent using comparison with a p-series

   B. The series is absolutely convergent using comparison with a geometric series

   C. The series is absolutely convergent using the Ratio Test.

   D. The series diverges.

☐ 1. $\sum_{n=1}^{\infty} \dfrac{\cos(8n)}{n!}$

☐ 3. $\sum_{n=1}^{\infty} \dfrac{(-1)^n \arctan n}{n^2}$

☐ 4. $\sum_{n=1}^{\infty} \dfrac{\sin(7n)}{3^n}$

Power Series

Section 8.5

Before Class Video Examples

1. Find the values of x for which the series will converge. (Give the interval and radius of convergence.)

   a. $f(x) = \sum_{n=1}^{\infty} \dfrac{(x-3)^n}{n}$

   b. $g(x) = \sum_{n=1}^{\infty} \dfrac{x^n}{n \cdot 3^n}$

   c. $h(x) = \sum_{n=1}^{\infty} \dfrac{4x^n}{n!}$

   d. $k(x) = \sum_{n=1}^{\infty} n!(3x-4)^n$

Section 8.5: Power Series

Algebra Review

1. **Absolute Value Inequalities**

$$|x| < C \implies -C < x < C$$

Example

i. Solve for x in the equation $|x - 2| < 10$

ii. Solve for x in the equation $|2x| < 1$

iii. Solve for x in the equation $|x^2| < 5$

iv. Solve for x in the equation $|-x^2| < 20$

v. Solve for x in the equation $\left|\dfrac{x^2}{7}\right| < 1$

A. Power Series

Definition: A power series is a series as a function.
$$f(x) = \sum_{n=0}^{\infty} C_n x^n = C_0 + C_1 x^1 + C_2 x^2 + C_3 x^3 + \cdots$$

Example: $g(x) = \sum_{n=1}^{\infty} x^{n-1} = 1 + x + x^2 + x^3 + \cdots$

Consider a geometric series as a function $f(x) = \sum_{n=1}^{\infty} a \cdot r^{n-1}$

Let $a = 1$ and $r = x$. To guarantee convergence, we must have $|x| < 1 \Rightarrow \sum_{n=1}^{\infty} x^{n-1} = \dfrac{1}{1-x}$

$\therefore f(x) = \sum_{n=1}^{\infty} x^{n-1} = 1 + x + x^2 + x^3 + \cdots = \dfrac{1}{1-x}$ if $|x| < 1$

Since $|x| < 1$ we have $|x| < 1 \Rightarrow -1 < x < 1$

Examples: Find the values for x for which the following series will converge. (Give the interval and radius of convergence.)

1. $f(x) = \sum_{n=1}^{\infty} \dfrac{x^n}{9^{n+1}}$

2. $g(x) = \sum_{n=1}^{\infty} \frac{(x-3)^n}{n}$

3. $h(x) = \sum_{n=1}^{\infty} \dfrac{(-1)^n (x+2)^n}{n \cdot 2^n}$

4. $k(x) = \sum_{n=1}^{\infty} n!(2x-1)^n$

5. $p(x) = \sum_{n=1}^{\infty} \dfrac{x^n}{n!}$

6. $f(x) = \sum_{n=1}^{\infty} \dfrac{1}{n^{2x+5}}$

WeBWorK

6. Find the interval and radius of convergence of the series $\sum_{n=1}^{\infty} \frac{(-9)^n x^n}{\sqrt[10]{n}}$

Representation of Functions as a Power Series

Section 8.6

Before Class Video Examples

1. Represent each function as a power series

   a. $f(x) = \dfrac{2}{1-x}$

   b. $g(x) = \dfrac{1}{1-3x}$

   c. $f(x) = \dfrac{1}{8-x}$

Section 8.6: Representation of Functions as a Power Series

2. Find the first four terms of the power series expansion

   a. $f(x) = \dfrac{3}{1+2x}$

   b. $f(x) = \dfrac{2}{1+x^2}$

Section 8.6: Representation of Functions as a Power Series 237

Algebra Review

1. **Fitting Standard Formulas**

Example

i. Write $5x - 2$ in the form $ax + b$. Give a and b.

ii. Write $\dfrac{2}{1-x}$ in the form $\dfrac{a}{1-r}$. Give a and r.

iii. Write $\dfrac{1}{1+x}$ in the form $\dfrac{a}{1-r}$. Give a and r.

iv. Write $\dfrac{2}{5+x^2}$ in the form $\dfrac{a}{1-r}$. Give a and r.

2. **Composition of Functions**

Products and Quotients of Functions

$$(f \cdot g)(x) = f(x) \cdot g(x) \qquad \left(\dfrac{f}{g}\right)(x) = \dfrac{f(x)}{g(x)}$$

v. Let $f(x) = 2\sqrt{x} + x$ and $g(x) = x$. Find the composition $f(x) \cdot g(x)$ and $\dfrac{f(x)}{g(x)}$.

Composite Functions

$$(f \circ g)(x) = f(g(x))$$

vi. Let $f(x) = x^2 + 2x$ and $g(x) = 2x$. Find the composition $f(g(x))$.

Section 8.6: Representation of Functions as a Power Series

A. Geometric Series

$$\sum_{n=1}^{\infty} ar^{n-1} = \sum_{n=0}^{\infty} ar^n \text{ converges to } \frac{a}{1-r} \quad \text{if} \quad |r| < 1$$

> Definition: A power series is a series as a function.
> $$f(x) = \sum_{n=0}^{\infty} C_n x^n = C_0 + C_1 x^1 + C_2 x^2 + C_3 x^3 + \cdots$$
>
> Example: $g(x) = \sum_{n=1}^{\infty} x^{n-1} = 1 + x + x^2 + x^3 + \cdots$

Examples: Represent each function as a power series

1. $f(x) = \dfrac{2}{1-x}$

2. $g(x) = \dfrac{1}{1+x^2}$

3. $h(x) = \dfrac{1}{2+x}$

B. Finding a Power Series Representation Using Derivation/Integration

Theorem: If $f(x) = C_0 + C_1(x-a)^1 + C_2(x-a)^2 + C_3(x-a)^3 + \cdots = \sum_{n=0}^{\infty} C_n(x-a)^n$

Then i. $f'(x) = 0 + C_1 + 2\cdot C_2(x-a)^1 + 3\cdot C_3(x-a)^2 + \cdots = \sum_{n=1}^{\infty} n\cdot C_n(x-a)^{n-1}$

ii. $\int f(x)\,dx = C_0(x-a)^1 + \dfrac{C_1}{2}(x-a)^2 + \dfrac{C_2}{3}(x-a)^3 + \cdots + C = \sum_{n=0}^{\infty} \dfrac{C_n}{n+1}(x-a)^{n+1} + C$

Examples: Represent each function as a power series

4. $f(x) = \tan^{-1} x$

Section 8.6: Representation of Functions as a Power Series

WeBWorK

4. Find a power series representation for the function $f(x) = \dfrac{x}{8+x^2}$. Give the first three nonzero terms.

5. Find a power series representation for $f(x) = \dfrac{16}{x^2 + 2x - 3}$.

$\dfrac{16}{x^2 + 2x - 3} = \dfrac{A}{x-1} + \dfrac{B}{x+3}$ where A = __4__ and B = __−4__

Find the first four nonzero terms in the power series representation of the following fractions:

$\dfrac{1}{x-1} = -1 - x - x^2 - x^3$

$\dfrac{1}{x+3} = \dfrac{1}{3} - \dfrac{x}{9} + \dfrac{x^2}{27} - \dfrac{x^3}{81}$

Therefore, $f(x) = \dfrac{16}{x^2 + 2x - 3} = c_0 + c_1 x + c_2 x^2 + \cdots$, where

$c_0 = -\dfrac{16}{3}$

$c_1 = -\dfrac{32}{9}$

$c_2 = -\dfrac{112}{27}$

$c_3 = -\dfrac{320}{81}$

242 Section 8.6: Representation of Functions as a Power Series

8. Evaluate the indefinite integral $\int \dfrac{2x - \tan^{-1}(2x)}{x^3}\, dx$ as a power series. Enter the first three nonzero terms in the power series representation of the following functions:

$$\dfrac{2x - \tan^{-1}(2x)}{x^3} = \dfrac{8}{3} - \dfrac{32}{5}x^2 + \dfrac{128}{7}x^4 - \cdots$$

$$\int \dfrac{2x - \tan^{-1}(2x)}{x^3}\, dx = C + \dfrac{8}{3}x - \dfrac{32}{15}x^3 + \dfrac{128}{35}x^5 - \cdots$$

The Radius of convergence = $\dfrac{1}{2}$

Section 8.6: Representation of Functions as a Power Series

9. Use a power series to approximate the definite integral $\int_0^{0.2} \dfrac{1}{1+x^6}\,dx$ to six decimal places. Enter the first four nonzero terms of the power series representation of the following functions:

$$\dfrac{1}{1+x^6} =$$

$$\int \dfrac{1}{1+x^6}\,dx = C +$$

Therefore, $\int_0^{0.2} \dfrac{1}{1+x^6}\,dx \approx$ \hspace{2em} (Correct to six decimal places)

Taylor and MacLauren Series

Section 8.7

Before Class Video Examples

1. Find the Taylor Series representation for $f(x) = \cos x$ centered at $a = \pi$. (Find the first three nonzero terms and the representation.)

2. Find the MacLaurin Series representation for $f(x) = e^x$. (Find the first four nonzero terms and the representation.)

Section 8.7: Taylor and MacLauren Series

3. Find the MacLaurin Series representation for $f(x) = e^{-5x}$. (Find the first four nonzero terms and the representation.)

4. Find the Taylor Series representation for $f(x) = \sin(\pi x)$ centered at $a = 0$. (Find the first three nonzero terms and the representation.)

A. Taylor Series

$$f(x) = f(a) + \frac{f'(a) \cdot (x-a)}{1!} + \frac{f''(a) \cdot (x-a)^2}{2!} + \frac{f^{(3)}(a) \cdot (x-a)^3}{3!} + \cdots = \sum_{n=0}^{\infty} \frac{f^{(n)}(a) \cdot (x-a)^n}{n!}$$

Example

1. Find the Taylor series representation for $f(x) = \cos x$ centered at $a = \dfrac{\pi}{2}$

$f(x) = \cos x \quad \rightarrow \quad f\left(\dfrac{\pi}{2}\right) = \cos\left(\dfrac{\pi}{2}\right) = 0$

$f'(x) = -\sin x \quad \rightarrow \quad f'\left(\dfrac{\pi}{2}\right) = -\sin\left(\dfrac{\pi}{2}\right) = -1$

$f''(x) = -\cos x \quad \rightarrow \quad f''\left(\dfrac{\pi}{2}\right) = -\cos\left(\dfrac{\pi}{2}\right) = 0$

$f'''(x) = \sin x \quad \rightarrow \quad f'''\left(\dfrac{\pi}{2}\right) = \sin\left(\dfrac{\pi}{2}\right) = 1$

$f^{(4)}(x) = \cos x \quad \rightarrow \quad f^{(4)}\left(\dfrac{\pi}{2}\right) = \cos\left(\dfrac{\pi}{2}\right) = 0$

$f^{(5)}(x) = -\sin x \quad \rightarrow \quad f^{(5)}\left(\dfrac{\pi}{2}\right) = -\sin\left(\dfrac{\pi}{2}\right) = -1$

$\vdots \qquad\qquad\qquad\qquad \vdots$

B. MacLauren Series

$$f(x) = f(0) + \frac{f'(0) \cdot (x)}{1!} + \frac{f''(0) \cdot (x)^2}{2!} + \frac{f^{(3)}(0) \cdot (x)^3}{3!} + \cdots = \sum_{n=0}^{\infty} \frac{f^{(n)}(0) \cdot (x)^n}{n!}$$

MacLauren Series is the specific case of Taylor Series centered at $a = 0$.

Examples

2. Find the MacLauren series representation for $f(x) = e^x$.

3. Find the MacLauren series representation for $g(x) = xe^x$.

C. Important MacLauren Series

- $\dfrac{1}{1-x} = \sum_{n=0}^{\infty} x^n = 1 + x + x^2 + x^3 + \cdots$

- $e^x = \sum_{n=0}^{\infty} \dfrac{x^n}{n!} = 1 + \dfrac{x}{1!} + \dfrac{x^2}{2!} + \dfrac{x^3}{3!} + \cdots$

- $\sin x = \sum_{n=0}^{\infty} \dfrac{(-1)^n \cdot x^{2n+1}}{(2n+1)!} = x - \dfrac{x^3}{3!} + \dfrac{x^5}{5!} - \dfrac{x^7}{7!} + \cdots$

- $\cos x = \sum_{n=0}^{\infty} \dfrac{(-1)^n \cdot x^{2n}}{(2n)!} = 1 - \dfrac{x^2}{2!} + \dfrac{x^4}{4!} - \dfrac{x^6}{6!} + \cdots$

- $\tan^{-1} x = \sum_{n=0}^{\infty} \dfrac{(-1)^n \cdot x^{2n+1}}{2n+1} = x - \dfrac{x^3}{3} + \dfrac{x^5}{5} - \dfrac{x^7}{7} + \cdots$

4. Find the MacLauren series representation for $h(x) = x \cdot \cos(2x)$.

Section 8.7: Taylor and MacLauren Series

D. Binomial Series

Combination Notation: $\binom{a}{b} = \dfrac{a!}{(a-b)! \cdot b!}$

Note: $\binom{k}{0} = \dfrac{k!}{0! \cdot k!} = 1$ and $\binom{k}{k} = \dfrac{k!}{k! \cdot 0!} = 1$

Example

5. $\binom{7}{3} =$

6. $\binom{5}{5} =$

Binomial Series

$$(1+x)^k = 1 + kx + \frac{k \cdot (k-1) \cdot x^2}{2!} + \frac{k \cdot (k-1) \cdot (k-2) \cdot x^3}{3!} + \cdots = \sum_{n=0}^{\infty} \binom{k}{n} \cdot x^n$$

where $k \in \mathbb{R}$ and $|x| < 1$

Example: Give a binomial series representation for each of the following:

7. $(1+x)^5$

8. $(1+x)^2$

The coefficients of each of these expansions can also be calculated using **Pascal's Triangle**

K
1
2
3
4
5
6
7

9. $(1+x)^7$

Section 8.7: Taylor and MacLauren Series

> **Binomial Series:** $(1+x)^k = 1 + kx + \dfrac{k \cdot (k-1) \cdot x^2}{2!} + \dfrac{k \cdot (k-1) \cdot (k-2) \cdot x^3}{3!} + \cdots$
>
> where $k \in \mathbb{R}$ and $|x| < 1$

Example: Give a binomial series representation for each of the following:

10. $\dfrac{1}{(1+x)^3}$

11. $\dfrac{2}{\sqrt{1-x}}$

WeBWorK

5. If $f^{(n)}(7) = \dfrac{(-1)^n n!}{3^n(n+2)}$ for $n = 0, 1, 2, \ldots$ then the Taylor series for f centered at 7 is

$$f(x) = \sum_{n=0}^{\infty} \underline{\hspace{2cm}} (x-7)^n$$

$$f(x) = \underline{\hspace{1.5cm}} + \underline{\hspace{1.5cm}}(x-7) + \underline{\hspace{1.5cm}}(x-7)^2 + \underline{\hspace{1.5cm}}(x-7)^3 + \cdots$$

6. Use the binomial series to expand the following function as a power series. Give the first three nonzero terms.

$$h(x) = \dfrac{1}{(3+x)^5} = \underline{\hspace{1.5cm}} + \underline{\hspace{1.5cm}} x + \underline{\hspace{1.5cm}} x^2 + \cdots$$

Section 8.7: Taylor and MacLauren Series

9. Use a Maclaurin series derived in this section to obtain the Maclaurin series for the given functions. Enter the first three nonzero terms only.

$f(x) = x\tan^{-1}(4x) =$

$f(x) = x^3 e^{-\frac{x}{2}} =$

12. Use series to evaluate the following limits. Give only two nonzero terms in the power series expansions.

$$\frac{1+4x-e^{4x}}{x^2} =$$

$$\lim_{x \to 0} \frac{1+4x-e^{4x}}{x^2} =$$

The Unit Circle

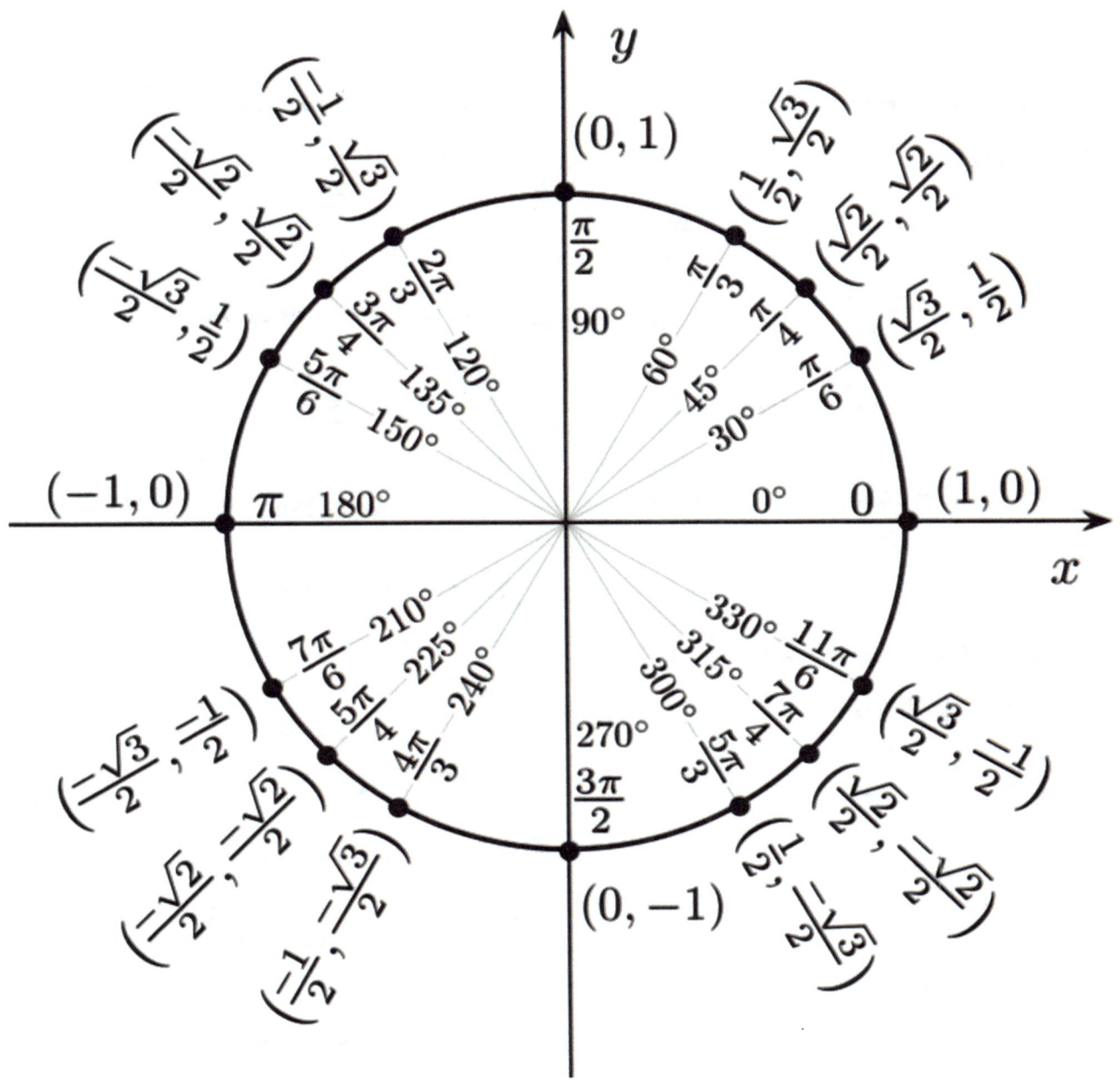

Trigonometric Identities

Trigonometric Functions

$\sin\theta = \dfrac{\text{opp}}{\text{hyp}} = \dfrac{y}{r}$

$\cos\theta = \dfrac{\text{adj}}{\text{hyp}} = \dfrac{x}{r}$

$\tan\theta = \dfrac{\text{opp}}{\text{adj}} = \dfrac{y}{x} = \dfrac{\sin\theta}{\cos\theta}$

$\csc\theta = \dfrac{\text{hyp}}{\text{opp}} = \dfrac{r}{y} = \dfrac{1}{\sin\theta}$

$\sec\theta = \dfrac{\text{hyp}}{\text{adj}} = \dfrac{r}{x} = \dfrac{1}{\cos\theta}$

$\cot\theta = \dfrac{\text{adj}}{\text{opp}} = \dfrac{x}{y} = \dfrac{1}{\tan\theta}$

Sum and Difference Formulas

$\sin(a \pm b) = \sin a \cos b \pm \cos a \sin b$

$\cos(a \pm b) = \cos a \cos b \mp \sin a \sin b$

$\tan(a \pm b) = \dfrac{\tan a \pm \tan b}{1 \mp \tan a \tan b}$

Double Angle Formulas

$\sin 2\theta = 2\sin\theta\cos\theta$

$\cos 2\theta = 1 - 2\sin^2\theta$

$\cos 2\theta = 2\cos^2\theta - 1$

$\cos 2\theta = \cos^2\theta - \sin^2\theta$

$\tan 2\theta = \dfrac{2\tan\theta}{1 - \tan^2\theta}$

Half Angle Formulas

$\sin^2\theta = \dfrac{1}{2}(1 - \cos 2\theta)$

$\sin\dfrac{\theta}{2} = \pm\sqrt{\dfrac{1 - \cos\theta}{2}}$

$\tan\dfrac{\theta}{2} = \pm\sqrt{\dfrac{1 - \cos\theta}{1 + \cos\theta}} = \dfrac{\sin\theta}{1 + \cos\theta} = \dfrac{1 - \cos\theta}{\sin\theta}$

$\cos^2\theta = \dfrac{1}{2}(1 + \cos 2\theta)$

$\cos\dfrac{\theta}{2} = \pm\sqrt{\dfrac{1 + \cos\theta}{2}}$

Pythagorean Identities

$\sin^2\theta + \cos^2\theta = 1$

$\tan^2\theta + 1 = \sec^2\theta$

$\cot^2\theta + 1 = \csc^2\theta$

Sum and Difference Formulas

$\sin a + \sin b = 2\sin\left(\dfrac{a+b}{2}\right)\cos\left(\dfrac{a-b}{2}\right)$

$\sin a - \sin b = 2\cos\left(\dfrac{a+b}{2}\right)\sin\left(\dfrac{a-b}{2}\right)$

$\cos a + \cos b = 2\cos\left(\dfrac{a+b}{2}\right)\cos\left(\dfrac{a-b}{2}\right)$

$\cos a - \cos b = -2\sin\left(\dfrac{a+b}{2}\right)\sin\left(\dfrac{a-b}{2}\right)$

Product Formulas

$\sin a \cos b = \dfrac{1}{2}[\sin(a+b) + \sin(a-b)]$

$\cos a \sin b = \dfrac{1}{2}[\sin(a+b) - \sin(a-b)]$

$\cos a \cos b = \dfrac{1}{2}[\cos(a+b) + \cos(a-b)]$

$\sin a \sin b = \dfrac{1}{2}[\cos(a-b) - \cos(a+b)]$

Law of Cosines

$a^2 = b^2 + c^2 - 2bc\cos A$

where A is the angle of a scalene triangle opposite side a.

Reduction Formulas

$\sin(-\theta) = -\sin\theta$

$\sin(\theta) = -\sin(\theta - \pi)$

$\tan(-\theta) = -\tan\theta$

$\mp\sin x = \cos(x \pm \tfrac{\pi}{2})$

$\cos(-\theta) = \cos\theta$

$\cos(\theta) = -\cos(\theta - \pi)$

$\tan(\theta) = \tan(\theta - \pi)$

$\pm\cos x = \sin(x \pm \tfrac{\pi}{2})$

Trigonometric Values for Common Angles

Degrees	Radians	sin θ	cos θ	tan θ	cot θ	sec θ	csc θ
0°	0	0	1	0	Undefined	1	Undefined
30°	π/6	1/2	√3/2	√3/3	√3	2√3/3	2
45°	π/4	√2/2	√2/2	1	1	√2	√2
60°	π/3	√3/2	1/2	√3	√3/3	2	2√3/3
90°	π/2	1	0	Undefined	0	Undefined	1
120°	2π/3	√3/2	-1/2	-√3	-√3/3	-2	2√3/3
135°	3π/4	√2/2	-√2/2	-1	-1	-√2	√2
150°	5π/6	1/2	-√3/2	-√3/3	-√3	-2√3/3	2
180°	π	0	-1	0	Undefined	-1	Undefined
210°	7π/6	-1/2	-√3/2	√3/3	√3	-2√3/3	-2
225°	5π/4	-√2/2	-√2/2	1	1	-√2	-√2
240°	4π/3	-√3/2	-1/2	√3	√3/3	-2	-2√3/3
270°	3π/2	-1	0	Undefined	0	Undefined	-1
300°	5π/3	-√3/2	1/2	-√3	-√3/3	2	-2√3/3
315°	7π/4	-√2/2	√2/2	-1	-1	√2	-√2
330°	11π/6	-1/2	√3/2	-√3/3	-√3	2√3/3	-2
360°	2π	0	1	0	Undefined	1	Undefined